THE SCIENTIFIC APPROACH TO EVOLUTION

THE SCIENTIFIC APPROACH TO EVOLUTION

What They Didn't Teach You in Biology

Rob Stadler

ISBN: 1532988095
ISBN 13: 9781532988097
Library of Congress Control Number: 2016907945
CreateSpace Independent Publishing Platform
North Charleston, South Carolina

The Six Criteria of High-Confidence Science

	Criteria of High-Confidence Science	Criteria of Low-Confidence Science
1	Repeatable	Not repeatable
2	Directly measurable and accurate results	Indirectly measured, extrapolated, or inaccurate results
3	Prospective, interventional study	Retrospective, observational study
4	Careful to avoid bias	Clear opportunities for bias
5	Careful to avoid assumptions	Many assumptions required
6	Sober judgment of results	Overstated confidence or scope of results

Contents

1 Evolution: The Great Divider · 1
2 The Six Criteria of High-Confidence Science · · · · · · · · · · · · 7
3 High-Confidence Science Could Save Your Life · · · · · · · · · 25
4 The Drinking Straw of Low-Confidence Science · · · · · · · · 33
5 Evolution Defined · 40
6 Low-Confidence Evidence for Evolution · · · · · · · · · · · · · · 46
7 High-Confidence Evidence for Evolution · · · · · · · · · · · · · · 81
8 Applying High-Confidence
  Evidence for Evolution · 119
9 It Is Called "Faith" · 136
10 The Scientific Higher Ground · 150

Appendix A   Hierarchy of Scientific Evidence · · · · · · · · · 159
Appendix B   Anticipated Objections · · · · · · · · · · · · · · · · 175
Bibliography · 185
Index · 193

1

Evolution: The Great Divider

The debate has been raging for more than 150 years, since the time of Charles Darwin. We have uncovered more than 100 times the fossil evidence that was known in Darwin's time; we have gained deep insight into genetics, molecular biology, and embryology; we have countless books by prominent intellectuals that endorse evolution as a fact; and in the United States, we have laws that evolution must be taught exclusively in public schools. And yet, the debate over the validity of evolution rages on.

According to repeated Gallup polls,[1] the proportion of Americans who believe "God created human beings pretty much in their present form at one time within the last 10,000 years or so" has hovered steadily in the mid–40 percent range for the last thirty years, with no clear trend. (See Figure 1, dotted line at the top.) I am left wondering how such a contentious issue could remain this stable for thirty years. Another 30-plus percent of Americans believe that "Human beings have developed over millions of years from less advanced forms of life, but God guided this process" (the dashed line in the middle). This result has also been relatively steady for thirty years, although it

---

[1] http://www.gallup.com/poll/21814/Evolution-Creation-Intelligent-Design.aspx.

appears to be decreasing over the last four years. The third and final survey option, "Human beings have developed over millions of years from less advanced forms of life, but God had no part in this process," has grown slightly in recent years, reaching 19 percent in the most recent survey (the solid line at the bottom). The survey results make it clear: as a nation, the United States is far from united when it comes to belief in evolution, and more than thirty years of intense pressure to teach evolution has had little impact.

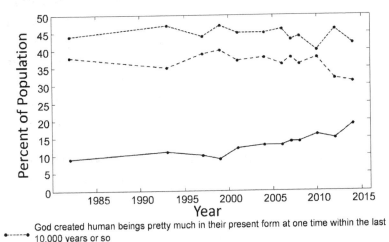

God created human beings pretty much in their present form at one time within the last 10,000 years or so

Human beings have developed over millions of years from less advanced forms of life, but God guided this process

Human beings have developed over millions of years from less advanced forms of life, but God had no part in this process

FIGURE 1: Results of thirty years of Gallup polls, showing American beliefs about human origins.

The Gallup poll categorizes all Americans into one of three world-views. Although there certainly is some variability of specific beliefs within each worldview, it is helpful to summarize the core beliefs of each worldview and each worldview's perceptions of the other worldviews.

For those with the "**God had no part in this process**" worldview, society's lack of acceptance of evolution provokes head-shaking, frustration, and disappointment. Despite great efforts to educate the public, little has changed in thirty years. Those who hold this view attribute this situation to religious fundamentalism or ignorance of science. They believe that evolution is a fact, proven by science, and that it should not be referred to as a "theory." The lack of acceptance of evolution may be viewed as an impediment to technological and economic progress. Those who do not accept evolution may be considered irrational, superstitious, simple-minded, and limited in their ability to contribute to society. Bill Nye (well known for his television role as "Bill Nye the Science Guy") posted a YouTube video called "Creationism Is Not Appropriate for Children" where he concludes:

> I say to the grown-ups: if you want to deny evolution and live in your world that is completely inconsistent with everything we observe in the universe, that's fine. But don't make your kids do it, because we need them. We need scientifically literate voters and taxpayers for the future. We need engineers that can build stuff and solve problems.[2]

The evolutionary biologist Richard Dawkins has a similar view, albeit expressed with more aggressive terminology: "If the history-deniers who doubt the fact of evolution are ignorant of biology, [then] those who think the world began less than ten thousand years ago are worse than ignorant, they are deluded to the point of perversity."[3]

---

[2] https://www.youtube.com/watch?v=gHbYJfwFgOU.
[3] Dawkins, R. The greatest show on Earth. New York: Free Press; 2009, p. 85.

These quotes can be viewed as outbursts of frustration over thirty years of stasis. Long-standing efforts to educate, with minimal effects, have degraded into the hurling of insults.

On the opposite extreme is the "**God created human beings**" worldview—those who refuse to accept evolution. Because they are committed to religious beliefs that God created the universe and life itself, evolution conflicts with their faith. They may not trust scientists who claim proof of evolution and may therefore grow to distrust the entire scientific establishment. They resent that public schools and media continuously barrage them with evolution, attempting to indoctrinate them and their children. If people accept that religious texts are wrong about creation, the credibility of the entire religious text will be questioned, and the foundation of their faith will crumble. They associate evolution with atheism, moral relativism, and general moral decline. The belief that humans are created in God's image is seen as an encouraging worldview, whereas belief that humans are animals and are a purposeless accumulation of random mistakes is a rather unfulfilling worldview. The "God had no part in this process" worldview offends their God, because evolutionists give God no credit for the incredible beauty of creation. Scientists and engineers are eager to receive credit for their discoveries and designs, and yet evolutionists refuse to give credit to the ultimate designer. Those who hold the "God created human beings" worldview will openly admit that what they believe requires faith, not scientific proof, and they argue that science does not support evolution. Ken Ham from the creationist group Answers in Genesis (and the Creation Museum in Kentucky) said during a debate with Bill Nye: "Now, at the Creation Museum, we're only too willing to admit [that] our beliefs [are] based on the Bible, but we also teach people the difference between beliefs and what one can actually observe and experiment with in the present."[4] With the words "we also teach people the difference between beliefs and what one can actually observe and experiment with in the present," Ham is suggesting that evolution cannot be observed and cannot be demonstrated experimentally.

Standing in the middle ground are the members of the "**God guided this process**" worldview—the 30-plus percent who attempt to combine evolution and faith in God. Exactly what role God played is a source of continual debate within this group. They argue that science and faith coexist without conflict, and they seek to provide interpretations of the religious texts that are fully compatible with evolution. They view the conflict between opposite extremes as unnecessary and detrimental. A central figure in this worldview is Francis Collins, leader of the Human Genome Project, author of *The Language of God: A Scientist Presents Evidence for Belief*, and founder of BioLogos. The mission of BioLogos is: "BioLogos invites the church and the world to see the harmony between science and biblical faith as we present an evolutionary understanding of God's creation."[5]

Given the Gallup poll results and this glimpse into the three opposing worldviews, a simple question arises: How can we ever resolve this conflict? The opposing views demonstrate that there are two foundations for determining truth: 1) science or 2) faith. The evolutionists who believe that God had no part in the process argue that their foundation is all science, without a hint of faith:

---

[4] Taken from the transcript of the Bill Nye / Ken Ham debate. February 4, 2014.
[5] http://biologos.org/about-us/our-mission/

Faith is the great cop-out, the great excuse to evade the need to think and evaluate evidence. Faith is belief in spite of, even because of, the lack of evidence.

—RICHARD DAWKINS

Conversely, those who refuse to accept evolution will admit that their belief is grounded in faith, not in science. This verse from the biblical book of Hebrews summarizes their view:

By faith we understand that the universe was formed at God's command, so that what is seen was not made out of what was visible.

—HEBREWS 11:3

The 30-plus percent of Americans who believe God somehow directed evolution think that they hold the solution to the conflict. Their solution is a type of compromise—relegating faith to those areas where science has not provided clarity.

This book offers a new approach to reconciliation. The approach starts with appreciation for different levels of confidence in scientific evidence and appreciation for the limitations of science (discussed in chapters 2 through 4). Subsequent chapters then review the evidence for evolution from this new perspective and reach profound conclusions. My hope is that the conflict over evolution can be lessened, at least to some extent, through this book.

2

The Six Criteria of High-Confidence Science

Science is a tool that is used to obtain knowledge. This tool has provided society with countless benefits: electricity, cures for diseases, airplanes, smartphones, and Cheetos, just to name a few. But, like any tool, science has to be used properly. Improper use can lead one to the wrong conclusion—the adoption of a fallacy. And like any tool, science has its limitations. A hammer has very limited value when trying to repair a wristwatch. Likewise, science has very limited value when trying to answer certain questions. It is important to be aware of these limitations when applying the tool of science. Again, failure to do so could lead to the wrong conclusion—the adoption of a fallacy.

Because science is a tool that is used to obtain knowledge, the application of the tool always follows an initial simple step: asking a question. Questions express a lack of knowledge. For some questions, the knowledge to answer them is already available: just ask Google. For other questions—a *subset* of those that have never been answered—the tool of science can be applied to provide an answer. Notice that I have emphasized the word "subset." There are plenty of unanswered questions for which science is not the right tool to provide an answer, such as: *What is the meaning of life? Should I ask her to marry me? Will the Packers win the Super Bowl?* Then there are questions where science is clearly the right tool, meaning that it can provide clear answers with high confidence: *How fast does a bowling ball fall? Does this drug reduce cholesterol levels? What is causing the decimation of bee populations?* And then there are questions where science may not be able to provide a confident answer, but we apply science because it is the best tool we have: *How did King Tut die? Does life exist on distant planets? Will next winter be warmer than average?* For these questions, science can only suggest lower-confidence answers.

In the paragraph above I presented an overly simplistic classification of questions into three categories:

a) those that science cannot answer;
b) those that it can answer with high confidence; and
c) those that it can attempt to answer, but only with low confidence.

I do not mean to give the false impression that every question falls distinctly into one of three bins. The level of confidence that science can provide is not black or white; there are many shades of gray with some subjectivity in assessing the degree of confidence. However, six simple criteria can be applied to assess the level of confidence. The following table compares the six criteria that separate high-confidence and low-confidence science. The remainder of this chapter explains why the six criteria are of such fundamental importance in the practice of science. Note that low-confidence science does not necessarily imply "bad" science. The assessment of low- or high-confidence science can follow from the applicability of the tool of science to address a question without implying a judgment of the skills or intentions of the scientist.

Criteria of High-Confidence Science	Criteria of Low-Confidence Science
1 Repeatable	Not repeatable
2 Directly measurable and accurate results	Indirectly measured, extrapolated, or inaccurate results
3 Prospective, interventional study	Retrospective, observational study
4 Careful to avoid bias	Clear opportunities for bias
5 Careful to avoid assumptions	Many assumptions required
6 Sober judgment of results	Overstated confidence or scope of results

1. REPEATABLE

Repeatable results are a hallmark of confident science. In applying science to answer a question, arriving at the same answer again and again is a means of building confidence. An experiment that hasn't been repeated (or cannot be repeated) either instills low confidence or is not credible. For this reason, scientists take very careful notes during experiments to make sure that all factors that can influence results are controlled and that the experiment can be repeated.

Let's imagine that I want to determine how fast a bowling ball falls, with a high degree of accuracy. I lift the ball up to a known height, start my stopwatch the moment I let go of the ball, and stop the watch exactly when the ball hits the ground. This is clearly an experiment that can be repeated again and again and again. To make the results repeatable and highly accurate, I have to control, or hold constant, many variables: the exact distance of the ball from the ground, the wind, the density of the air, and the location of the experiment. (The force of gravity varies by about 0.7 percent at various locations on the earth.) I also have to make sure that the ball is not moving before it is released. If I succeed in controlling all factors that could influence the results, the result should be very repeatable and would have been obtained with high confidence.

Conversely, let's ponder the question: How fast did a bowling ball fall five thousand years ago? Quite likely, this experiment was conducted five thousand years ago, albeit in a much less formal and controlled way, with a rock representing the bowling ball (reminiscent of a *Flintstones* episode). Unfortunately, ancient measurement instruments were not very precise, and we don't have access to the methods or the results. We can all make educated guesses at the answer to this question, and your guess may well be correct, but without the aid of a time machine, there is no way to repeat this experiment to provide a high-confidence answer. Thus, any attempt to answer this question will rely heavily on assumptions and the bias of the investigator. We will discuss bias and assumptions later; for now, the simple message is that **high-confidence science must be repeatable**.

2. DIRECTLY MEASURABLE AND ACCURATE RESULTS

The word "accurate" has an obvious association with high-confidence science. In my career as a scientist, every piece of measurement equipment that I use, whether it is measurement of time, voltage, weight, pressure, or temperature, must have one thing in common: a valid calibration sticker. Each piece of equipment is calibrated by comparing its result to a known standard to ensure accuracy. If the equipment passes this calibration step, it receives a sticker of certification. If I use equipment that is not properly calibrated, I have to take corrective action by repeating the experiment with calibrated equipment.

As an example from my childhood, I frequently joined the neighborhood kids for backyard football. The quarterback on offense was matched against a "rusher" on defense. The rusher kept the quarterback from running the ball and put pressure on him to pass the ball quickly. But the rusher couldn't just chase down the quarterback at the start of the play—the rusher had to be delayed to give the quarterback some time to pass the ball. We started out requiring the rusher to count to five—slowly—before going after the quarterback. But we found that some kids counted to five faster than others. In other words, the measurement of time was not accurate because the rushers were not properly calibrated. We then attempted to improve the accuracy by adding a new rule: the rusher had to say "banana" after counting each number. If you ever enter a laboratory and find a scientist measuring time by saying "1 banana, 2 banana, 3 banana..." it is safe to classify this as low-confidence science. Properly calibrated, accurate measurements are required for high-confidence science.

Direct measurement of results also helps to provide high confidence. In the experiment on falling bowling balls, the results were presented directly on my handy stopwatch. If the watch has been calibrated for accuracy and is used properly, this provides highly confident results. Many applications of science do not allow such direct measurements: the more indirect a measurement becomes, the less confidence it provides.

Doctors commonly ask: "What is the patient's blood pressure?" This refers to the pressure inside an artery. Despite decades of research on the importance of blood pressure and on the technology of measuring blood pressure, the one and only direct measurement of blood pressure is to poke a hole in an artery and insert a catheter or a pressure sensor. Not only is this painful, but it carries a small risk of significant bleeding, infection, or blood-clot formation. For this reason, we generally prefer an indirect measurement of blood pressure. A cuff around the arm applies increasing pressure until the pressure outside of the brachial artery exceeds the blood pressure inside the artery. The cuff pressure is then slowly released while listening with a stethoscope applied over the artery. When the cuff pressure decreases to below the systolic pressure (the larger number, such as 120 mmHg in the commonly quoted "120 over 80" normal blood pressure), the artery opens and closes with each heartbeat, producing a sound known as a Korotkoff sound. When the pressure in the cuff decreases below the diastolic pressure (the smaller number, or 80 mmHg in the commonly quoted "120 over 80" normal blood pressure), the Korotkoff sounds can no longer be heard.

The resulting indirect estimate of blood pressure lacks the accuracy of direct catheterization of an artery. Inaccuracies may result from clothing, use of the wrong-size cuff, patients who boast about their grandchildren while the nurse is trying to hear the Korotkoff sounds, or from applying the cuff above or below the level of the heart. Although these limitations are well known, the practice of medicine makes a compromise: preferring the convenient, indirect measurements of cuff pressure over the inconvenient, direct measurements of blood pressure with a catheter.

Our previous question, "How fast did a bowling ball fall five thousand years ago?" is an extreme example of indirect measurement. In this case, nobody was there to witness the experiment or to provide an accurate result to us in the present time. Thus, to address this question, most people would conduct the measurement today and assume that nothing has changed in the last five thousand years. This is a very indirect measurement. We will discuss how assumptions degrade the confidence of scientific results later. For now, remember that **directly measured and accurate results are important for high-confidence science.**

3. PROSPECTIVE, INTERVENTIONAL STUDY

The best way to use science to obtain new knowledge is to design a study in advance (the "prospective" part) where one or more variables are manipulated (the "interventional" part) and all other variables are controlled so that they do not confound the results. Our bowling ball experiment was designed in advance; therefore, it is prospective. The study manipulated one variable (dropping the ball from a known distance) and controlled other predetermined variables such as location on the earth, wind speed, etc., so it is a prospective interventional study. The opposite of prospective study design is retrospective study, where one considers something that happened in the past. The opposite of an interventional study is an observational study, where you passively observe what occurs or has occurred. Thus, a retrospective observational study is the opposite extreme of a prospective interventional study.

In the study of bowling ball behavior, it is possible to imagine a retrospective observational experiment to address the same question. I could review thousands of hours of video recordings of people interacting with bowling balls. In this footage, there certainly would be cases where bowling balls are dropped. I suspect several of these clips have already ended up on YouTube. From these clips I could estimate the distance and the duration of the drop. It would thus be possible to estimate how fast a bowling ball falls. But clearly, this would provide a lower-confidence answer. The reason for the lower confidence is that I am not able to control all of the confounding variables in the retrospective analysis: Was the ball dropped from a resting state, or was it pushed? Exactly how high was the ball when it was dropped? What was the geographic location of the ball? What was the density of the air at that time and place? In addition, the video recordings were probably obtained at thirty or sixty frames per second, so time measurement is not very accurate.

Here is one of the most important points in this book: **By controlling confounding variables, well-conducted prospective interventional experiments are able to *conclude causality*, that is, determine what caused the results that were observed. Retrospective observational studies are not able to control variables and therefore can only *suggest associations*, not conclude causality.**

In my prospective experiment on bowling balls, where I controlled all confounding variables, I can conclude that gravity alone caused the observed result. In the retrospective review of videos, I can observe that letting go of the bowling ball is associated with the ball moving toward the floor at a specific rate of acceleration. But I cannot establish a cause-and-effect relationship—I cannot confidently conclude that gravity as we know it was the cause of the bowling ball's acceleration. What if the video had been recorded on the moon, where gravity is about one-sixth as strong as it is on Earth? What if Photoshop was used to manipulate the video? What if the video recording occurred in a "fun house" room that was built upside down, so that the floor was on the ceiling, and the person holding the ball and the camera were strapped upside down to the ceiling? When they release the ball, it would appear to move toward the ceiling, not the floor. I realize that these are highly unlikely scenarios, but it's also highly unlikely that I would conduct an accurate study of bowling balls in the first place. The point is, retrospective observational studies provide lower-confidence results because they can't provide assurance of control over confounding variables. Thus, they cannot *conclude causality*; they can only *suggest associations*.

4. AVOIDING BIAS

Everyone has bias. Everyone. Bias is like a pair of glasses that you look through as you make observations. Your bias affects the way that you perceive everything. It is a sinister agent: we all have it, but we all like to believe that we don't have it. Bias can come from your goals, your desires, your education, your experiences, your upbringing, your religious beliefs (or lack thereof)—or it can come from outside pressures. Bias is devastating to science. Two people can be presented with the exact same experimental results and come up with completely different interpretations. When you get to know the people, what they are passionate about, and what they are trying to achieve, you can usually understand their interpretation in light of their bias. In my career as a scientist, the power of bias to cloud the truth has always amazed me and has often humbled me. Every time something I worked on had a disappointing performance, I could trace that disappointment back to a bias that I'd had in the research or an incorrect assumption that was made along the way.

High-confidence science, and good scientists, must make every effort to exclude bias. One cannot simply attempt to act in an unbiased manner—one must actively practice habits of excluding bias. Any clinical research protocol that I write must have a section called "Minimization of Bias," where I describe the procedures that will be followed to exclude bias. All of the other criteria of high-confidence science, to some degree, are ways of avoiding bias. Repeating experiments is a way of reducing bias. Directly measured results avoid the biases inherent in indirect results. Conducting prospective interventional experiments is a way of reducing bias over retrospective or observational studies. Assumptions are made, and found to be acceptable, because of bias. Finally, bias is extremely difficult to suppress when results are being interpreted or discussed.

The US Food and Drug Administration (FDA) has this to say about bias: "Any clinical trial may be subject to unanticipated, undetected, systematic biases. These biases may operate despite the best intentions of sponsors and investigators, and may lead to flawed conclusions. In addition, some investigators may bring conscious biases to evaluations."[6]

To explain how bias can operate despite the best intentions of an investigator, imagine that Fred wanted to answer the question: What percentage of human beings currently have cancer? He decides to answer this question in a socially awkward way: he walks up to everyone whom he meets and asks them if they currently have cancer. After gathering data from one thousand people, he summarizes his results into a conclusion: 52.5 percent of all people currently have cancer.

You probably are thinking that this percentage is far too high. But Fred can show you his raw data—even video recordings of the one thousand interviews—and prove that his result is accurate. Now do you believe the result? You suspect that Fred's results are somehow biased, so you ask him more questions. Fred is offended that you are

---

[6] US Department of Health and Human Services, Food and Drug Administration. Guidance for industry: Providing clinical evidence of effectiveness of human drug and biological products. May 1998. http://www.fda.gov/downloads/drugs/guidancecomplian ceregulatoryinformation/guidances/ucm072008.pdf.

accusing him of bias. After all, he just asked a simple question and accurately recorded the results. You eventually find out that he works in a hospital: in the cancer clinic, to be specific. It's no wonder that the majority of people whom he encountered currently have cancer. His results have what is called sampling bias. This bias was not a conscious effort and did not occur because Fred had some external pressure, religious belief, or hidden agenda. He simply asked a group of people who were not representative of the general population but he presented his results as if they *were* representative of the general population.

Other forms of bias are not as innocent. As stated in the FDA quote, some investigators may bring conscious biases to evaluations. Have you ever noticed that fortune cookies are biased? Almost without exception, each fortune cookie contains an encouraging message or a positive outlook. If fortune cookies were unbiased, some would contain encouraging messages and some would predict IRS audits, broken relationships, or the development of embarrassing rashes. The reason for this is clear: the people who make the fortune cookies and the restaurants who serve them want your business. Sales would certainly decrease over time if customers paid good money for discouraging messages. This is a conscious form of bias. Unfortunately, the practice of science commonly struggles with similar biases. Positive study results bring fame, publications, and more grant money for academicians. Positive results bring revenue for industry. These pressures subtly encourage researchers to explain away negative results, give up on publishing negative results, or find a way to put a positive spin on what they find. I don't want to imply that scientists can't be trusted, but we do have to keep in mind that scientists are human, and all humans have bias. **It is hard to overestimate the power of bias to corrupt science, even when the best intentions are at work.**

5. AVOIDING ASSUMPTIONS

Assumptions are shortcuts or ways of filling in for something that is unknown. Because well-conducted scientific studies are expensive, time consuming, and often tedious, shortcuts become extremely attractive. We discussed earlier that high-confidence experiments should control the important confounding variables. Controlling variables takes effort and may not always be possible. In my bowling ball experiment, for example, the density of air will have a very minor impact on the acceleration of the ball. Controlling air density is rather difficult. To do it right, the experiment should be conducted in an environment where air pressure and composition are controlled. As a result, I might be willing to take a shortcut by assuming that the impact of air density on the acceleration of the ball is negligible. Having made this assumption, it is very important that I follow up with two actions. First, I need to acknowledge this assumption in my description of the experiment. Second, I need to include a justification for making the assumption, such as a calculation of the impact of a range of air densities on the acceleration of a ball.

The second type of assumption, that of filling in for something that is unknown, often occurs after the experiment. Imagine that I conducted the bowling ball experiment and properly controlled all of the variables that we have discussed. After completing the experiment and presenting my results, my colleague asks what kind of polish I applied to the bowling ball. She says that a rough surface creates more wind drag than a well-polished surface. I don't have the ball anymore, and I didn't pay attention to how the ball was polished. Rather than repeat the entire experiment, I am eager to convince her to assume that the polish had a negligible impact.

Although avoiding assumptions increases the confidence of the result, almost every study includes at least a few assumptions.

Acknowledging the assumptions and providing a justification is important. Disregarding them, usually because one is convinced by his or her own bias that the assumptions are justifiable, is likely to degrade the quality of the science. **Undocumented assumptions— either because you have justified them in your own mind or because you unconsciously made them—can lead to the adoption of fallacy.**

6. SOBER JUDGMENT OF RESULTS

A write-up of scientific findings culminates in the "Discussion" and "Conclusion" sections of a manuscript. Here, the authors summarize their findings and the implications of what they found. We all know that humans are hungry for success, for recognition, for impact. As a result, we all have a bias to amplify the scope and the importance of what we have found. **Good scientists will suppress this bias by providing a sober judgment of their results, claiming that the results are to be expected only under the specific conditions of the experiment.** Poor scientists will amplify their results, claiming that what they found applies well beyond the scope of their experiment and has far-reaching implications.

During prospective studies, the result is obtained while controlling any confounding variables. The result then applies only under those specific controls and not if the confounding variables had different values. My experiment on bowling balls occurred at a specific geographic location, with a specific distance of drop, a specific air density, and specific wind conditions. My conclusion from the study about bowling ball behavior strictly applies only under the conditions that I studied. It would be irresponsible for me to conclude that the same behavior would occur in a broader set of conditions: for example, during a tornado, or if the experiment were conducted at the top of Mount Everest or on the moon. With retrospective or observational studies, the situation worsens. Again, we cannot assume that the same results would be obtained in a different situation, but with retrospective or observational studies we often are unsure exactly what situation produced the results. The confounding variables are not well controlled, or we are not aware of how they were controlled. How, then, can we provide proper scope for the application of the results?

For those who are weary of bowling balls and hunger for a more practical example, aspirin is a well-known drug for decreasing the risk of blood clotting. The biostatistician Robert Glynn and colleagues took on the task of determining if daily small doses of aspirin would reduce blood clots in healthy women.[7] The study that they conducted was immense: 39,876 women were randomly selected to receive either aspirin or placebo (i.e., a sugar pill) and were followed for ten years. After conducting such a large and lengthy study, their conclusion was: "These data suggest that long-term, low-dose aspirin treatment has little effect on the prevention of VTE in initially healthy women."

Here, "VTE" stands for venous thromboembolism, a medical term that translates to: blood clots that form in veins and travel toward the heart to cause trouble. The concluding statement from these authors provides a sober judgment of what they found. They could have selected different words for their conclusions; for example: "Our results prove that long-term aspirin treatment has no effect on VTE."

Let's look at the difference between these two concluding sentences. After studying nearly forty thousand women for ten years,

---

[7]   Glynn, RJ, et al. Effect of low-dose aspirin on the occurrence of venous thromboembolism. Ann Intern Med. 2007;147:525-33.

they chose to use the words "suggest" and "little effect." After such a large study, it seems that they would have earned the use of the words "prove" and "no effect." The words "suggest" and "little effect" are an important part of their sober judgment of results.

Linguists call this "hedging."[8] Hedging has two purposes: it keeps readers from over-interpreting the results and it allows authors to express the extent of uncertainty about the importance and validity of the results. By choosing the word "suggest," Glynn and colleagues are admitting that other studies with differing approaches could produce different results. Perhaps if the study were conducted only on Japanese women, or post-menopausal women, or women who frequently consume Cheetos, the results would be different. By choosing "little effect" they are admitting that an even larger study may detect a very small benefit, because their study may not have been large enough to detect such a small benefit.

They also specifically mention "low-dose" aspirin and "initially healthy women" in their conclusion, to frame the scope of their study. My alternate conclusion makes it seem that the results could apply to any dose aspirin and could be equally valid for men, healthy women, or unhealthy women. Thus, my version of the conclusion artificially expands the scope of the actual study, which is irresponsible.

One easy way to spot overstated confidence or scope of results is to look for words that imply extremes, like *always*, *never*, or *optimal*. These words suggest a conversion of the author from scientist to salesperson. While these words rarely reflect reality, they are commonly used to compensate for weak or low-confidence results. On the other hand, the scientist may provide a sober judgment of his

---

[8] Hyland, K. Hedging in scientific research articles. Amsterdam and Philadelphia: John Benjamins Publication Company; 1998. DOI: 10.1075/pbns.54.

or her results, but others' subsequent reviews of the findings may amplify the results.

Another way to provide a sober judgment of results is to include a candid discussion of study limitations. This practice is becoming the norm among medical journals. Milo Puhan, a professor of epidemiology and public health at the University of Zurich, found that 73 percent of 300 medical research papers included a frank discussion of study limitations.[9] He summarizes the importance of being open about study limitations (because *all* studies have limitations):

> This will not only benefit science but also offers incentives for authors: If not all important limitations are acknowledged, readers and reviewers of scientific articles may perceive that the authors were unaware of them. Authors should take advantage of their content knowledge and familiarity with the study to prevent misinterpretations of the limitations by reviewers and readers.

Hopefully the scientific community will agree with these words of wisdom, and all fields of science will adopt an expectation of the open disclosure of study limitations. If you think that I'm nuts for making such a suggestion, allow me to point out that even nuts now provide a sober judgment of their benefits. Figure 2 is from a can of almonds. The label says, "Scientific evidence *suggests, but does not prove,* that eating 1.5 ounces per day of most nuts, such as almonds, as part of a diet low in saturated fat and cholesterol *may* reduce the risk of heart disease." (Emphasis on hedging terms added.)

---

[9] Puhan, MA, et al. Discussing study limitations in reports of biomedical studies—The need for more transparency. Health and Quality of Life Outcomes. 2012;10:23.

FIGURE 2: Even nuts have to be careful to provide a sober judgment of results. This can says, "Scientific evidence suggests, but does not prove, that eating 1.5 ounces per day of most nuts, such as almonds, as part of a diet low in saturated fat and cholesterol may reduce the risk of heart disease."

CHAPTER SUMMARY

Science is a tool applied to answer questions. The tool of science can provide a high-confidence answer to a question if the question can be addressed in a manner that is repeatable, is directly measured and accurately measured, is prospectively studied by applying an intervention and controlling confounding factors, is free from bias, is free from assumptions, and is interpreted with sober judgment. These six criteria of high-confidence science are fundamental tenets of science—no field of science is exempt from their reach. To argue for an exemption, one would have to argue that higher confidence arises if a scientific result cannot be repeated; or if indirect measurements, rather than direct measurements, are conducted; or if retrospective observational study (with a lack of control over confounding variables), rather than prospective interventional study, is conducted; or if bias or assumptions are generously applied; or if the results are claimed to apply to broader or different situations than the actual scope of the study.

For those who question the fundamental importance of these six criteria or for those who desire further description of the applicability of the six criteria, please take the time now to read Appendix A. This appendix shows how the six criteria have been adopted in the practice of science, with particular emphasis on the practice of medicine. Because medicine is a life-or-death application of science, medicine has developed a particularly keen appreciation for the six criteria of high-confidence science.

3

High-Confidence Science Could Save Your Life

In Western cultures, ischemic heart disease (a.k.a. heart attacks) is the leading cause of death, and high cholesterol is a leading cause of ischemic heart disease. As a result, the last twenty-five years have been an all-out war against cholesterol. The "big guns" in this war include a family of drugs called "statins" (atorvastatin, simvastatin, lovastatin, pravastatin, and rosuvastatin). Chances are that you or a close friend are taking one, because current medical guidelines suggest that a whopping 17.5 percent of the entire population of the United States should be taking one, including 50 percent of people between the ages of forty and seventy-five.

With so many lives at stake, it would be comforting to know that treatments for high cholesterol are grounded in high-confidence science. To increase our familiarity with the six criteria of high-confidence science, let's consider whether the question "Is there a drug that can significantly reduce cholesterol?" can be addressed by high-confidence science. This will serve as a useful comparison as we consider the level of confidence associated with the evidence for evolution. The following table summarizes the six criteria for this question.

Criteria of High-Confidence Science	How They Apply To The Question: Is there a drug that can significantly reduce cholesterol?
1. Repeatable	Thousands of clinical trials have been conducted.
2. Directly measurable and accurate results	Blood cholesterol can be measured directly by an automated and properly calibrated machine.
3. Prospective, interventional study	It is easy to design a prospective drug experiment. Administering a cholesterol-lowering drug is a type of intervention.
4. Careful to avoid bias	Bias can be reduced by randomly assigning patients to receive the drug or a placebo (sugar pill), and by "double-blinded" study design (meaning that neither the patient nor the doctor know if the patient is receiving the cholesterol pill or the sugar pill. A separate group of researchers keeps track of this).
5. Careful to avoid assumptions	Any assumptions in each specific study should be stated and justified.
6. Sober judgment of results	This will be up to individual authors who write up the experiments. Their scientific integrity will be demonstrated if they only make claims that are well supported by the data.

We conclude that the question, "Is there a drug that can significantly reduce cholesterol?" can be addressed by high-confidence science. This is not to say that every study *will* contain high-confidence science, but certainly this question *can* be addressed by high-confidence science. Now, let's zoom in to explore one particular study that provides an exemplary high-confidence result.

In the middle of 2015, the FDA approved even bigger weapons in the war against cholesterol: alirocumab and evolocumab. These two drugs are prescribed for those who continue to have high cholesterol after taking maximum doses of statins. As an aside, these drugs are actually antibodies that are produced by hordes of genetically modified Chinese hamster ovary cells. Because the drugs are antibodies that would be destroyed in the stomach, they can't be taken in pill form—they have to be injected under the skin. These drugs are expected to generate about $10 billion in revenue per year, and the current value of one ounce of these drugs is around $210,000, or about 175 times the value of gold—not bad for a bunch of mutant Chinese hamster ovary cells.

A very large clinical study was prospectively designed and conducted to answer the question: For patients on maximum statin therapy, does the addition of alirocumab cause a significant reduction of cholesterol? The study was called the "ODYSSEY LONG TERM" trial, and the results are available in *The New England Journal of Medicine*.[10] The online supplementary materials include all 1,623 pages of the clinical protocol and 76 pages of supplementary appendix. They enrolled 2,341 patients who had already maxed out on statin therapy but still had high cholesterol. The patients were randomized to receive biweekly injections of alirocumab or placebo. (Yes, the placebo group received thirty-nine injections of nothing but saline!) Neither the patients nor the medical staff knew which treatment the patients received. Separate research staff had to keep track of which treatment was given to each patient; this type of study is called "double blinded". The patients were studied for 1.5 years. The primary result was the change in LDL cholesterol (the "bad cholesterol") between the start of the study and the first twenty-four weeks of treatment. The LDL cholesterol was measured in a centralized facility, rather than at each individual clinical site, and the results were not presented to the clini-

---

[10] Robinson, JG, et al. Safety and efficacy of alirocumab in reducing lipids and cardiovascular events. N Engl J Med. 2015;372:1489–99.

cal sites. A secondary result included careful tracking of cardiovascular events (like heart attacks and strokes) between the time of the first injection and up to ten weeks after the last injection. Any reported adverse events had to be reviewed by an independent committee to judge the validity and the severity of the events. Patients who decided to discontinue the injections were still followed, and their cholesterol readings and other results were still included. This is called "intention to treat" analysis and is a way to avoid bias when subjects drop out of a study. The authors concluded:

> In conclusion, in the ODYSSEY LONG TERM trial, 2341 high-risk patients were randomly assigned to either the PCSK9 inhibitor alirocumab or placebo. As compared with placebo, alirocumab reduced LDL cholesterol levels by 62 percentage points at 24 weeks, with a consistent reduction over a period of 78 weeks of treatment. In a post hoc analysis, there was evidence of a reduction in cardiovascular events with alirocumab.

Looking at these results, I'd like to emphasize the way in which the authors framed the scope of their findings. The drug was studied among high-risk patients, and the authors make that clear. The measurement of effect occurred at twenty-four weeks of treatment, and they make that clear. This clarity provides a sober judgment of results. The authors are not attempting to extrapolate the results beyond the scope of their study, and their modest claims keep the reader from over-interpreting the results: for example, by assuming that the drug is also beneficial for low-risk patients or that the drug will reduce LDL cholesterol levels by 62 percentage points over many years of treatment. Finally, they also present the second efficacy result, "evidence of a reduction of cardiovascular events," with proper scope. The report does not say something like "conclusive proof" or "convincing evidence" or "absolute certainty"; it humbly reports the result as more of a hint: "evidence of…" Also, the second efficacy result was not prespecified, meaning that the authors did not plan to conduct this comparison in advance. They properly labelled it a "post hoc analysis." After data is collected, scientists can perform any number of unplanned analyses in what we might consider a type of fishing expedition, trying to find associations between the data that they collected and the clinical result. For example, they could divide their 2,341 patients up, looking for benefits among different subgroups, such as women, men, smokers, diabetics, or people who consume significant quantities of Cheetos. But each statistical test that is performed increases the chances that one test will turn up as being statistically significant simply by random chance. To avoid this, the most impressive studies prespecify the statistical tests that they will perform *before* the first data point is obtained. In this study, the authors were very straightforward in pointing out the limitation of their post hoc finding and downplaying their confidence in this particular result—a real sign of scientific integrity.

Finally, the authors included a paragraph with a summary of the study limitations. One of the noted limitations was that the study duration was relatively short for treatment of a chronic disease. Another stated limitation was that "the number of cardiovascular events was relatively small, which limits the robustness of these data and the confidence that they are not simply a chance finding." Given that their study showed a statistically significant reduction in cardiovascular events with alirocumab treatment, admitting to such a limitation is a good sign of transparency and scientific integrity.

The table below summarizes the features of this study that make it an exemplary high-confidence scientific study.

Criteria of High-Confidence Science	Conduct of the ODYSSEY LONG TERM trial
1. Repeatable	The study can be repeated. A very similar study was conducted on evolocumab, with similar results.
2. Directly measurable and accurate results	LDL cholesterol levels were measured by an automated machine in a centralized facility.
3. Prospective, interventional study	• The study was designed in advance of the first subject enrolment. • The intervention was administration of alirocumab. • Randomization was applied to control confounding variables. • The study was registered in a national database before it began (see Appendix A). • The primary goal of the study was pre-specified. • Analyses that were not pre-specified were labelled "post hoc" and were listed as limitations.
4. Careful to avoid bias	• The study was placebo-controlled. • Patients were randomized to drug or placebo. • The study was double blinded. • The study was conducted at multiple centers to dilute any center bias. • They performed "intention to treat" analysis. • All occurrences of cardiac events were reviewed by an independent adjudication committee.
5. Careful to avoid assumptions	Having studied the manuscript carefully, I was not able to find any assumptions.
6. Sober judgment of results	• Study limitations were openly discussed. • The study was conducted on "high risk" patients. Conclusions state that the drug worked in "high risk" patients, not in a broader population. • Post hoc analyses were stated as such, and the observed statistically significant reduction in cardiovascular events with alirocumab was downplayed because of small numbers. • Results included an indication of variability. • Statistical tests concluded that the results were highly unlikely to result from random chance.

This manuscript stands as a great example of how to produce high-confidence scientific results.

Here is a similar question that could be answered by high-confidence science: Is the drug digitalis beneficial for heart failure or atrial fibrillation (a common heart arrhythmia)? The foxglove plant is used to produce digitalis (digoxin). This drug has been used for over two hundred years as a treatment for heart failure and more recently as a treatment for atrial fibrillation. Both of these uses of digitalis have been endorsed by clinical practice guidelines, as established by professional societies of physicians. Clearly, this drug *could* be studied with very similar methods as the cholesterol-lowering drugs and the same level of confidence could be achieved. Yet, despite being endorsed by clinical practice guidelines, only one randomized controlled clinical trial has been conducted (the "Digitalis Investigation Group" or "DIG" trial[11]) for the heart failure indication, and no randomized controlled clinical trials have been conducted for the atrial fibrillation indication. The DIG trial studied 6,800 patients for three years and found that digitalis and placebo had similar rates of death, but digitalis reduced the number of hospitalization events for heart failure. Many other studies have been conducted, but they were all lower-confidence designs (registries, retrospective studies, post hoc analyses, and observational studies). As a result, the evidence of the benefit of digitalis remains weak.

A pair of recent reports—both of which summarized results across multiple studies that included over 300,000 patients—suggest that the use of digitalis actually *increases* mortality (i.e., causes more deaths)

---

[11] Garg, R, et al. The effect of digoxin on mortality and morbidity in patients with heart failure. N Engl J Med. 1997;336:525–33.

among heart failure and atrial fibrillation patients.[12,13] As a result, even after studying hundreds of thousands of patients, the use of digitalis for heart failure and atrial fibrillation remains controversial. Why? Not because the question cannot be addressed by high-confidence science, but because the evidence-gathering approaches have been weak.

Why are cholesterol-lowering drugs so well studied, while digitalis is not? Large, randomized, controlled clinical studies typically cost $10–50 million to conduct. Because digitalis has been around for two hundred years, it is not possible to patent it or to make a large profit from it. If there is not much money to be made from it, there is not much money to conduct research on it. As a result, nobody has been willing to pay for decisive high-confidence clinical trials on the use of digitalis.

---

[12] Wang, ZQ, et al. Digoxin is associated with increased all-cause mortality in patients with atrial fibrillation regardless of concomitant heart failure: A meta-analysis. J Cardiovasc Pharmacol. 2015;66:270–5.

[13] Vamos, M, et al. Digoxin-associated mortality: A systematic review and meta-analysis of the literature. Eur Heart J. 2015;36:1831–8.

4

The Drinking Straw of Low-Confidence Science

Here's an experiment for you. Go to McDonald's and get yourself a free drinking straw. Then take it to a football game. You are to keep one eye closed and use the other eye to look only through the straw. Your job is to be the referee on the field, keeping the game fair so that the best team wins. At the same time, many cameras will capture every aspect of the game. At the end of the game, we will compare your job as a referee to the "truth" as captured by the cameras and see if you did a good job.

In a scientific endeavor, obtaining the truth can feel a lot like looking through a drinking straw, especially if the question that you are trying to answer can only be addressed with low confidence. Make no mistakes, even the questions that can be addressed with high confidence, like "Is digitalis beneficial for heart failure or atrial fibrillation?" are quite challenging to answer. In this chapter, we will consider an example question that can only be addressed by low-confidence science. Here, the question is so difficult that it may never be answered, and there is real risk of accepting an incorrect answer.

The example question is: How did King Tutankhamun die? We believe that he died at age nineteen, during the ninth year of his reign. Since the discovery of his tomb in 1922, debate has raged over the cause of his premature death. Before we discuss any theories, let's assess the situation. The most important point to raise is that we can only conduct retrospective observational science, not a prospective experiment. As we discussed in chapter 2, only prospective experimentation is able to control confounding variables in order to demonstrate cause and effect. For our question, the effect is death, and the cause is what we are trying to determine. Retrospective or observational studies can only suggest association. Therefore, all we can hope to achieve is a list of observations that *may* be associated with his death. The associations may strongly support a particular mode of death, or they may not. Subjectivity will exist over the level of support for a particular mode of death, thus opening the door for bias. The table below further assesses how our question stacks up against the criteria of high-confidence science.

Early examinations of the mummy, including x-ray images collected in 1968, suggested signs of trauma to the back of the head and an unhealed broken leg.[14] This led to suggestions that King Tutankhamun died of trauma such as being in a chariot accident, being kicked by a horse, or being attacked by a hippopotamus (no, I did not make that one up). Others suggested that he was killed by a blow to the head or that he was poisoned. Some have hypothesized that his death was a result of familial disease. Artwork that depicts King Tut and his father Akhenaton tends to have a markedly feminized appearance, which could suggest genetic disorders such as Marfan syndrome, Wilson-Turner X-linked mental retardation syndrome, or androgen insensitivity syndrome. A simple Google search on this topic reveals quite a list of publications and theories but no signs of a convergence.

---

[14] Harrison, RG. Postmortem on two pharaohs: Was Tutankhamun's skull fractured? Buried Hist. 1971;4:114–29.

Criteria of High-Confidence Science	How they apply to "How did King Tut die?"
1. Repeatable	"Repeatable" here refers to the death of King Tut. It is not possible to repeat this.
2. Directly measurable and accurate results	Quite likely someone witnessed his death, but we don't have a detailed description and they didn't have a good understanding of modern medicine.
3. Prospective, interventional study	Any evidence that we could dig up (literally) is retrospective and observational. Causality cannot be determined by retrospective observational studies.
4. Careful to avoid bias	Results in the form of simple observations can be objective, such as: the mummy has a fractured knee. Subjectivity cannot be avoided when combining the observations into a hypothesized cause of death. Subjectivity opens the door for bias.
5. Careful to avoid assumptions	Because we have no direct account of the death, assumptions are unavoidable.
6. Sober judgment of results	Beware of theories with falsely exaggerated claims or confidence.

In 2010, a new study was published—one that brought advanced technology to bear.[15] Let's zoom in to explore this study in detail. I need to be clear up-front that I am not questioning the integrity of the authors and I am not claiming that they are bad scientists. They are

---

[15]   Hawass, Z, et al. Ancestry and pathology in King Tutankhamun's family. JAMA. 2010;303:638–47.

honestly seeking the truth, but they have selected an extremely challenging area of study. Like refereeing a football game while looking through a drinking straw, it is not easy to determine what actually happened.

The scientists conducted a study on sixteen royal mummies, including King Tutankhamun. They collected computed tomographic imaging (CT scans) of the mummies, conducted genetic analysis on DNA extracted from each mummy, and combined this data with the known anthropological findings. From the CT scans, the authors focused on malformation of the feet and suggested that Tutankhamun had a walking disability. The DNA analysis showed no evidence of familial inherited disorders, but they found DNA from the parasite that causes malaria. Although malaria can be fatal, evidence of malarial infection certainly does not infer cause of death. Here, we can appreciate that observations cannot determine a cause-and-effect relationship, they can only suggest associations. In response to this publication in the Journal of the American Medical Association, letters to the journal editor conveyed skepticism over the DNA analysis, noting that DNA may not be able to survive for 3,300 years and that DNA contamination is very difficult to avoid. Their study concluded:

> Tutankhamun had multiple disorders, and some of them might have reached the cumulative character of an inflammatory, immune-suppressive—and thus weakening—syndrome. He might be envisioned as a young but frail king who needed canes to walk because of the bone-necrotic and sometimes painful Köhler disease II, plus oligodactyly (hypophalangism) in the right foot and clubfoot on the left. A sudden leg fracture possibly introduced by a fall might have resulted in a life-threatening condition when a malaria infection occurred.

Seeds, fruits, and leaves found in the tomb, and possibly used as medical treatment, support this diagnosis.

To their credit, the authors admit a lack of confidence by their choice of hedging words such as "possibly" and "might." The table below shows how this specific study lines up well with the criteria of low-confidence science.

Some people may remain hopeful that science will eventually provide a confident explanation for King Tut's death. However, appreciating the lack of clarity from the studies thus far, and recognizing that retrospective observational studies can only suggest associations, not demonstrate cause-and-effect, makes it highly unlikely that science will ever provide a confident explanation. New evidence may arise, and new technology can be brought to bear, but we will never know for sure how King Tut died.

This example shows the difficulty that the tool of science can have in addressing certain questions. The limitations of science are too great to arrive at a high-confidence answer. In this case, the great age of the subject of interest is the main source of the limitations. In general, the limitations of science become more apparent as one travels farther back in time, because the available evidence becomes increasingly indirect. For example, I would suggest that science could provide a higher-confidence answer to the question "How did Michael Jackson die?" a lower-confidence answer to the question "How did King Tut die?" and only a very-low-confidence answer to the question "How did this fossilized dinosaur die?"

As a reminder, although we cited this example as low-confidence science and the cholesterol-lowering drugs as high-confidence science, remember that these classifications need not be black and white; a wide spectrum of gentle gradations exists between high-confidence and low-confidence science.

Criteria of Low-Confidence Science	Conduct of the Study on King Tut's Family
1. Can't be repeated	Although the data collection and analysis can be repeated, the fundamental question is: How did King Tut die? The circumstances that led to his death cannot be repeated.
2. Indirectly measured, extrapolated, or inaccurate results	Very likely someone witnessed his death, but we can't interview them and they didn't provide documentation. Even if a witness documented the cause of death, it may not make sense in light of modern medicine. The available evidence (DNA, bone malformations, leg fracture, seeds and fruit found in the tomb) indirectly relate to the cause of death.
3. Retrospective, observational study	This study involves observations about the distant past. The results can suggest associations but cannot determine a cause-and-effect relationship.
4. Clear opportunities for bias	Because the results are subjective, one person's opinion against another, the interpretation of the observed findings is highly dependent on personal bias.
5. Many assumptions required	Here are a few of the assumptions behind the findings. Some may have strong archaeological support and others may not: • The body is actually that of King Tut • The body was not damaged by the burial or excavation process • The malaria was associated with his death • DNA can survive 3300 years, and contamination did not occur
6. Overstated confidence or scope of results	To their credit, the authors admit a lack of confidence in their conclusion.

It is important to note that **if a question cannot be addressed by high-confidence science, the results from the best available scientific approach should not be portrayed as being high-confidence results.** For those disciplines in which high-confidence science *can* be practiced, the results of high-confidence science should be prioritized over lower-confidence results. For those disciplines in which only lower-confidence science can be practiced, the results should be appropriately stated with candid acknowledgement of lower confidence, which is the best that can be expected. The real problem occurs when low-confidence science is inappropriately portrayed as high-confidence science.

Our work now comes to the focus of this book: the assessment of the confidence level of scientific evidence of evolution. We will see that it is possible to practice high-confidence science when studying evolution, and several high-confidence studies have been conducted. However, much of the evidence that is commonly presented is very low confidence. Before we jump into assessing the evidence, we must first define a few terms.

5

Evolution Defined

A typical heated debate about evolution consists of two sides that passionately disagree on every point, yet both sides *assume* that they agree on one foundational point: the definition of "evolution." In reality, they often don't even agree on the definition of evolution. They are not aware of this discrepancy, however, which then confuses and fuels the argument. Bearing this in mind, it's best to start with clear definitions of terms.

The most **generalized definition of evolution** is: changes in the properties of organisms that occur over more than one lifetime. This definition is generalized because any change that occurs, big or small, beneficial or detrimental, can be called evolution. Because small changes are directly observable—such as the great variety of dog breeds that all derived from wolves, or bacteria that have developed resistance to antibiotics—this generalized form of evolution is irrefutable. If any readers are feeling resistance here, I hope it is directed at this definition of evolution, and not at the claim that changes in the properties of organisms can be directly observed.

Perhaps the most commonly assumed definition of evolution involves the grandest extent of the concept. We will call this "**grand evolution**," which is the theory that all life proceeded from a common ancestor through the slow accumulation of changes as a result of random mutations and natural selection. This grand definition of evolution is what is taught in our classrooms, by law, and is the subject of the surveys mentioned in chapter 1. Grand evolution implies generalized evolution (i.e., major evolutionary changes imply the occurrence of minor evolutionary changes), but generalized evolution does not imply grand evolution (i.e., minor evolutionary changes do not imply major evolutionary changes). It might help to refer to Figure 3.

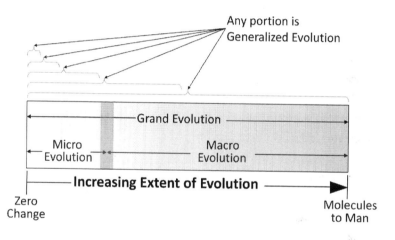

FIGURE 3: A diagrammatic representation of the different scales of evolution.

Grand evolution can be divided into two mutually exclusive subsets called "microevolution" and "macroevolution." **Microevolution** includes only minor changes, like the variety of dog breeds and microbes that become resistant to antibiotics. **Macroevolution** describes larger changes, such as lizards evolving into birds. It is very hard to define exactly where microevolution ends and macroevolution begins. Starting with the 18th century Swedish botanist Carl Linnaeus, scientists have organized all forms of life into a structure consisting of taxonomic levels. You may be familiar with the lowest taxonomic levels: genus and species (humans, for example, are of the genus *Homo* and the species *sapiens*). The common definition of microevolution in current biology books includes evolutionary changes below the species level. Thus, microevolution becomes macroevolution when one species branches into two distinct species, a process called "speciation." According to this definition, the direct observation of speciation (as presented in chapter 7) would be proof for macroevolution and could thus imply proof of grand evolution; this is because the two components of grand evolution, microevolution and macroevolution, are both directly observable. To make the boundary between microevolution and macroevolution more relevant to the debate at hand, we will define microevolution as minor evolutionary changes, *including* the process of speciation. Macroevolution then includes the portion of grand evolution that is too large to be called microevolution.

The table below and Figure 3 summarize these four definitions of evolution.

Generalized evolution	Changes in the properties of organisms that occur over more than one lifetime
Grand evolution	The theory that all life proceeded from a common ancestor through slow accumulation of changes as a result of random mutations and natural selection
Microevolution	Minor evolutionary changes, including the process of speciation
Macroevolution	The portion of grand evolution that includes changes too large to be called microevolution

Those who do not accept the grand definition of evolution argue that a very important distinction should be made between microevolution and macroevolution. They accept microevolution as the full extent of evolution, thus rejecting macroevolution.

In contrast, those who believe in the grand definition of evolution do not see an important distinction between microevolution and macroevolution, because they view these to be the same process. They perceive that evidence in support of microevolution is equally valid as evidence in support of macroevolution. Claims such as "Evolution is supported by an overwhelming amount of scientific evidence" are commonly backed by examples of microevolution, like microbial resistance to antibiotics and finches with different beaks. For example, in *The Greatest Show on Earth*, Richard Dawkins offers as evidence for evolution the various breeds of dogs (all evolved from wolves) and the various forms of cabbage (broccoli, cauliflower, kale, brussels sprouts, etc.), which all evolved from wild cabbage. This is microevolution. But he then concludes, "if so much evolutionary change can be achieved in just a few centuries or even decades, just think what might be achieved in ten or a hundred million years," which refers to macroevolution.[16]

Perhaps unexpectedly, those who believe in the grand definition of evolution commonly assume that you either accept all of it or none of it, probably because they don't see an important distinction between microevolution and macroevolution. As such, those who reject the grand definition of evolution are commonly assumed to reject microevolution as well. Bill Nye said, "It's not just that they don't understand how evolution led to the ancient dinosaurs, for example, they take it another step and deny that evolution happened at all anywhere, let alone that it is happening today."[17]

---

[16]   Dawkins, R. The greatest show on Earth. New York: Free Press; 2009, p. 37.

[17]   Nye, B. Undeniable: Evolution and the science of creation. New York: Saint Martin's Press; 2014, p. 10.

Because of this assumption, and the fact that small evolutionary changes over generations have practical implications for biology and medicine, Nye is outraged by those who do not accept the grand definition. The misconception that you either accept all of it or none of it led Nye to a rather extreme viewpoint:

> Inherent in this rejection of evolution is the idea that your curiosity about the world is misplaced and your common sense is wrong. This attack on reason is an attack on all of us. Children who accept this ludicrous perspective will find themselves opposed to progress. They will become society's burdens rather than its producers, a prospect that I find very troubling.[17]

If you deny that small changes occur over generations, then you probably are limiting the contributions you can make to the biological sciences. But Nye needn't be so concerned. Even the most conservative creationist groups, like Answers in Genesis, openly agree that speciation occurs. They believe speciation is a result of the genetic variability built into creation. They believe that Noah's Ark contained only a small fraction of the number of species that we see today—a necessity for fitting the animals on the Ark. Speciation occurred after the Flood to expand the number of species to what we observe today. As said by Ken Ham: "when it comes to finches, we actually would agree as creationists that different finch species came from a common ancestor, but a finch is what that would have to come from."[18] Therefore, it certainly is possible, and is quite common, to deny the grand definition of evolution but accept microevolution, including speciation. Perhaps the "all or none" misconception occurs in part because groups like Answers in Genesis don't like to use the term "evolu-

---

[18] Taken from a transcript of the Bill Nye / Ken Ham debate, February 4, 2014.

tion" when they speak of such speciation. This is because 1) this could be misinterpreted as acceptance of the grand definition of evolution and 2) "evolution" implies an advancement or gain in genetic information, whereas they generally view speciation as a *decrease* in genetic information.

Hopefully this discussion helps to clarify the various definitions of evolution and emphasizes the importance of defining terms before entering a debate. People who deny the grand definition of evolution would be less likely to end up in heated arguments if they started by agreeing that microevolution does occur. By taking this position, they will dramatically reduce the scope of the argument, because evidence of microevolution is mutually accepted—it is no longer relevant to the debate. Conversely, people who accept the grand definition of evolution should not assume that acceptance is an "all or none" phenomenon; they certainly should not assume that people who reject the grand definition are opposed to progress and are a burden to society.

Our discussion will therefore be focused on the grand definition of evolution and macroevolution. We will now proceed to assess the confidence level of the evidence for evolution. We will start by discussing the evidence that only provides low confidence and will eventually get to the high-confidence evidence.

6

Low-Confidence Evidence for Evolution

This chapter will show that much of what is reported as evidence of macroevolution does not fare well on the scales of high-confidence science. For those who feel their defenses rising, take heart, because chapter 7 will discuss the evidence for evolution that is high confidence. I also want to reiterate that low-confidence science does not imply "bad science" or "intentionally misleading science." Scientists in these fields have taken on incredibly difficult challenges in trying to apply the tool of science within areas where science has very limited power to uncover the truth.

THE FOSSIL RECORD

The fossil record is the most debated evidence for macroevolution. Both sides of the debate see the fossil record as supporting their side and contradicting the opposing view. Those who believe in the grand definition of evolution suggest that the fossil layers are clearly organized from simple forms of life (deep layers) to complex forms of life (shallow layers). For example, you never see a mammal fossilized in a layer of sediment from the Devonian period. They are quick to point out series of fossils that look like transitions from one type of life to another. For example, ungulates (terrestrial mammals) could appear to transition to cetaceans (e.g., whales) via the series of fossilized organisms known as pakicetus, rodhocetus, and dorudon. On the other side, those who do not believe in the grand definition of evolution suggest that if Darwin was right, transitional forms should be found everywhere in the fossil record and not just in rare and overly hyped examples. Rather than transitional forms, the fossil record speaks of discontinuities such as the "Cambrian explosion," which shows evidence of the sudden appearance of almost all phyla of life in a relatively short period of time. They argue that the layers of millions of fossils found all over the earth and the polystrate fossils (i.e., upright tree trunks that extend through multiple geological strata) are evidence of a global flood.

We'd like to know if the tool of science can address this question with high confidence: "Were the life-forms of the fossil record produced by macroevolution?" Stepping back for a moment to consider this question, we are asking if the sparse remnants of organisms that died long ago can provide confident evidence of the process that brought the organisms into existence. This seems to be a very indirect way to search for the truth—quite different than giving someone a drug and observing a decrease in cholesterol levels. In the discussion of King Tut, we concluded that the farther back in time you go, the more difficult it is to determine a cause of death. But here, we are going well beyond the difficulty of determining cause of death: we are trying to determine the process that caused the (now fossilized) life-forms to come into existence. This makes refereeing a football game while looking through a drinking straw sound easy. Let's consider each of the six criteria of high-confidence science in turn.

1. Repeatable. Although the process of finding fossils can be repeated, that does not address our question. The repeated finding of a fossil would address a question like: "Can a fossil of this type be found?" Our question is about what caused the organisms to come into existence. As such, "repeatable" here refers to repeating the process and the conditions that produced the organisms found in the fossil record. Just as the death of King Tut cannot be repeated, neither can the process that produced the organisms found in the fossil record. Not even the movie *Jurassic Park* could accomplish this. In the movie, a few dinosaurs were brought back to life through a type of cloning process; I've never heard anyone theorize that a type of cloning process produced the organisms of the fossil record. Repeatability requires setting up the same conditions and observing the result again and again and again. Only a time machine could allow us to know the conditions and to observe the process that produced the life-forms that have been fossilized.

2. Directly measurable and accurate results. The fossil record itself is directly measurable, but again, that does not address our question. Direct measurements of the fossil record would address questions like: "Do fossils exist?" or "Where are they found?" or "How large was the femur of a T. rex?" We're interested in the process that produced the life-forms that are preserved in the fossil record. The direct measurement of this process would require examination of the birth of thousands of generations of life-forms that existed a long time ago. Until we get that time machine, this is clearly not possible. What we observe by finding fossils is very, very indirect evidence of the process that produced those life-forms. A current biology textbook has this to say: "The most direct evidence that evolution has occurred is found in the fossil record."[19]

I could not disagree more.

---

[19] Raven, PH, et al. Biology. Eighth edition. New York, NY: McGraw-Hill, Inc.; 2008.

3. Prospective, interventional study. Observations of the fossil record are the antithesis of prospective, interventional studies—they are retrospective observations to the extreme. Remember that prospective, interventional studies are required to demonstrate causality, and our question is concerned with the *cause* of the life-forms seen in the fossil record. With retrospective observations, we can only hope to *suggest association*. Associations are easy to make when we look back into a pile of fossils, but those associations may not be meaningful.

Let me explain further with an example (see Figure 4). Imagine a very large number of fossil skulls from a large variety of vertebrates. The skulls are organized according to the sedimentary fossil layer where they were found: skulls from deeper layers at the bottom and skulls from shallower layers at the top. Perhaps there would be thousands of skulls in each layer. Given all these organized fossils, we pick a random skull from the deepest layer that contains a vertebrate skull (e.g., Skull 2 from Layer 1 in Figure 4) as a starting point and pick a random skull from the shallowest layer as an ending point (e.g., Skull 999 from Layer 100 in Figure 4). We then find a volunteer and ask he or she to select a series of skulls, one selected from each layer, that will make the best step-by-step progression from the appearance of the starting skull to the appearance of the ending skull. It takes a while, but the person does a great job selecting the best choice of skulls that appear to progress from the start to the end (the bold rectangles in Figure 4). We could then pull out those selected skulls and lay them out as a series (bottom panel of Figure 4). Looking at this associated series of skulls, one might get the impression that the progression of appearance is stunning evidence for macroevolution. Yet we have no evidence whatsoever of causality: no evidence that macroevolution is the cause of the apparent progression from skulls in deeper layers to skulls from shallower layers. All we have is a simple association of shapes as fossil layers progress from deep to shallow. This will, I hope, help to clarify the difference in confidence levels afforded by a prospective interventional study as opposed to a retrospective observational study. To put this concept into practical terms, biology textbooks commonly cite the aforementioned pakicetus, rodhocetus, and dorudon as steps to whale evolution from terrestrial mammals. How can these three retrospective observational pieces of evidence possibly provide high confidence for the process that is responsible for producing whales? All that they can do is suggest an association: a set of fossils that appear to show a progression of characteristics.

4. Careful to avoid bias. There are two major reasons why bias is pervasive in addressing this question. First, the fossil record can only provide sparse and very indirect evidence of the process that produced these life-forms. Although we can make objective statements about fossils (e.g., where they were found, how deep, how large, how heavy), any claims about how the fossilized life-forms came to exist are inherently subjective, because we have no hope of repeating the process, no direct measurements of the process that created them, and no hope of establishing causality. Figure 5 illustrates a useful metaphor. When the level of confidence in the evidence is high, the opportunity for bias is low. Conversely (and very much true in this case), when the level of confidence in the evidence is low, the opportunity for bias is high. In this case, the influence of bias can easily overpower the influence of evidence in determining the truth.

Second, the question we are trying to address is highly contentious and polarizing. Emotions run high, and opinions are very strong. The selected quotations in this book should give a feel for the high emotional level of this topic. This situation is the opposite of the seesaw metaphor of Figure 5: high emotions lead to high bias, while low emotions lead to low bias. Dr. Spock from *Star Trek* did not have emotions, which made him a highly valued, unbiased science officer (although not so skilled in the area of romance!).

Table of fossilized vertebrate skulls

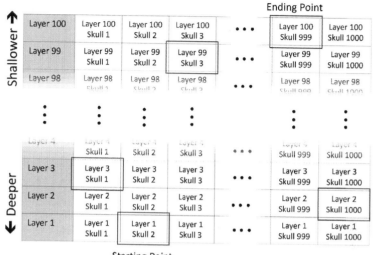

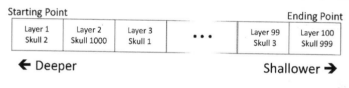

FIGURE 4: A table containing one thousand fossilized skulls per sedimentary layer, with deeper layers at the bottom and shallower layers at the top. The bold rectangles indicate skulls that are selected, one per layer, because they show a progression of appearance from the starting point (Layer 1, Skull 2) to the ending point (Layer 100, Skull 999). When arranged as a sequence (bottom panel), they may give the impression of an evolutionary process, but they can only suggest association; they cannot demonstrate causality.

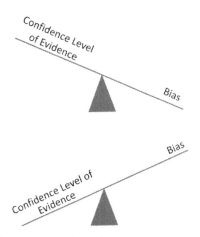

FIGURE 5: When the level of confidence in the evidence is high, the opportunity for bias is low. Conversely, when the level of confidence in the evidence is low, the opportunity for bias is high.

One of the world's foremost paleontologists, Jack Horner, presented a brilliant TED talk in November 2011 entitled "Where Are the Baby Dinosaurs?"[20] He presented numerous impressive dinosaur finds, each classified as a separate species and bearing the name selected by its discoverer. He then pondered why we never find baby versions of the named dinosaurs. By cutting into the skull bones and performing microscopic analysis of bone structure, he was able to estimate the maturity of each animal at death. As a result, many separate dinosaur species turned out to be different levels of maturity of the same dinosaur species. Infants, adolescents, and adults of the same dinosaur species had been incorrectly classified as separate species. Horner explained that this occurred because "scientists have egos, and

---

[20] https://www.ted.com/talks/jack_horner_shape_shifting_dinosaurs?language=en.

scientists like to name things" in order to obtain fame in the discovery of a new species. This is a simple example of how bias can be a very powerful force whose influence can overwhelm the limited power of the tool of science in a field such as this. When addressing a hotly contested issue like evolution, the power of bias will be further amplified. If bias could lead paleontologists to adopting the fallacy that a single fossilized species was actually several separate fossilized species, then a similar bias could lead paleontologists to adopt a fallacy that several distinct life-forms with no evolutionary linkage represent an evolutionary sequence of life-forms.

5. Careful to avoid assumptions. Assumptions must be applied when something is unknowable but is an important part of an argument. Combining various pieces of evidence from the fossil record into a story of what produced the fossilized life-forms inherently requires many assumptions, because we cannot directly observe how the life-forms came to be. To this end, the seesaw metaphor of Figure 5 applies to assumptions as well: a high level of confidence in the evidence implies little room for assumptions, whereas low confidence in the evidence implies a great need for assumptions. A few of the common assumptions involved in interpreting the fossil record are as follows.

a. A series of bones from different layers of sediment that appears to show a progression in appearance (e.g., skulls with increasing brain size or lengthening jawbone) implies that the life-forms of the deeper layer of sediment evolved into the life-forms of the shallower layer of sediment.

b. A supernatural creator would only create clearly distinct life-forms. The creator would not have created a series of similar life-forms that could be misinterpreted as an evolutionary sequence.

c. The absence of a particular organism in a fossil layer implies that that organism did not exist at that time. For example, the absence of fossil remains of mammals in the Devonian strata implies that no mammals existed when these layers were formed.

d. The absence of a predicted series of transitional fossils in the fossil record does not imply that the transitional life-forms never existed. For example, the assumption that the absence of transitional forms leading up to the Cambrian explosion (or Cambrian radiation) occurred because the soft tissues of the numerous transitional forms were not fossilized.

(Note that the last two assumptions are contradictory, yet both are frequently made when presenting fossil evidence in favor of macroevolution.)

6. Sober judgment of results. In light of the discussion of the previous five criteria, a sober scientific judgment would be to state upfront that one cannot hope to use the tool of science to provide a confident answer to what produced the life-forms we see in the fossil record. This is not an antiscience view or a pessimistic view or a harshly critical view; rather, it is the only sober scientific conclusion. It is important to note that this is not a conclusion that favors one side of the debate over the other: neither side can claim that the fossil record strongly supports their argument.

The table below summarizes how addressing the question "Were the life-forms of the fossil record produced by macroevolution?" is an extreme form of low-confidence science.

Low-Confidence Evidence for Evolution

Criteria of Low-Confidence Science	How they apply to the question: Were the life-forms of the fossil record produced by macroevolution?
1. Can't be repeated	The conditions that produced the life-forms seen in the fossil record cannot be repeated.
2. Indirectly measured, extrapolated, or inaccurate results	Direct measurement of the process that produced the life-forms found in the fossil record is impossible. Studying fossils is a very indirect means of studying the process that produced the life-forms.
3. Retrospective, observational study	Finding fossils and using them as evidence is an extreme form of retrospective observational study. The results can suggest associations but cannot determine a cause-and-effect relationship. The question that we are trying to answer is about cause. It cannot be answered by retrospective observational data.
4. Clear opportunities for bias	Because the evidence is so indirect, interpretation of the observed findings is highly dependent on personal bias. The bias is amplified by the strong polarization of people on opposite sides of the debate.
5. Many assumptions required	The synthesis of collections of observations about fossils into a story of how the fossilized life-forms were produced requires many assumptions to fill in the unknowns.
6. Overstated confidence or scope of results	Study of the fossil record involves scant evidence from very distant history. Given the inherent uncertainties, upfront statements of assumptions, prolific use of hedging language, and candid presentation of study limitations should be expected. It would be most appropriate to state upfront that one cannot hope to use science to provide a confident answer.

The level of confidence that the FDA requires in order to sell a new drug or device provides a sobering comparison here. The use of fossil evidence to determine what process is responsible for producing life-forms is an *extreme* of low-confidence science, and it is time that we put this evidence in its proper place. The tool of science is being stretched well beyond its limitations, which carries a high risk for the adoption of fallacy.

Please understand that I am not arguing that paleontologists are bad scientists or that they are intentionally misleading anyone. I am saying that, like refereeing a football game while looking through a drinking straw, trying to explain how the life-forms found in the fossil record came to be is an extremely difficult problem. They are stretching the tool of science beyond its limits. The result is inherently very low confidence, regardless of the great efforts, intelligence, and good intentions of paleontologists.

Modern Organisms

Perhaps our confidence level will increase if we consider the macroevolution of modern (extant) organisms from the fossil record. We could consider a question such as: Did modern life-forms emerge from the life-forms of the fossil record by evolutionary processes? A good example here is the evolution of humans from other primates. Note that this does not include cases of observable microevolution (e.g., observable speciation, breeds of dogs, or microbial resistance). Those cases will be discussed in the next chapter on high-confidence evidence of evolution.

Because modern organisms are directly observable, this question feels more approachable and less subjective than our prior question on the fossil record. But we are still relying on the fossil record—sparse remnants of dead organisms—to explain the process of producing new kinds of organisms. Although we can directly observe modern life, we cannot directly observe the process that produced that life. This question thus faces the same challenges as discerning the process that produced the life-forms in the fossil record:

- The process that created modern life-forms cannot be repeated.
- Although we can conduct direct measurements on modern life-forms, nobody witnessed the process that produced these life-forms.
- This is a retrospective observational study, not a prospective or interventional study. It cannot establish cause and effect.
- Because the evidence is so indirect, any interpretation of the observed findings is highly dependent on personal bias; the bias is amplified by the strong polarization of people on opposite sides of the debate.
- The synthesis of collections of observations about fossils into a story of how modern organisms came to be requires many assumptions to fill in the many unknowns.
- Given the inherent uncertainties involved, upfront statements of assumptions, the prolific use of hedging language, and candid presentation of study limitations should be expected.

Let's dig a little deeper (no pun intended) into a specific case: the macroevolution of humans from primates.

Lucy

On November 24, 2015, a special Google Doodle (artwork based on the Google brand and displayed on the Google homepage) celebrated the forty-first anniversary of the discovery of Lucy.[21] Lucy is the

---

[21] http://www.google.com/doodles/41st-anniversary-of-the-discovery-of-lucy.

most complete specimen of what has been classified as the species *Australopithecus afarensis*. Lucy is not the only fossil of *Australopithecus afarensis*, nor is she the first to be found, but she is the most famous. Because it is easier to say "Lucy" than *Australopithecus afarensis*, I'm going to refer to both the individual specimen and to this collection of fossils as "Lucy." The Google Doodle contained a series of three figures—on the left, a figure that looked like a chimpanzee, ambulating on all four limbs; on the right, a human walking upright; and in the middle, Lucy walking on two legs with stature and appearance intermediate between the chimpanzee and human. The Google Doodle clearly implied that Lucy existed within the direct evolutionary lineage to humans. This is certainly a common image representing grand evolution. But is it grounded in high-confidence science?

Donald Johanson, the discoverer of Lucy, wrote a book called *Lucy: The Beginnings of Humankind.* The book provides great insights into the work of paleoanthropologists, including the challenges that they face and their struggle to obtain confident results. Johanson's candid perspective speaks to the analogy of attempting to referee a football game while looking through a drinking straw. He admits that defining a separate species from fossils is a very subjective process (recall the above discussion of the baby dinosaur TED talk by Jack Horner):

> Somewhere along the evolutionary path that leads from one to the other, a species line may be drawn if, in the opinion of anatomists, the differences between them are significant. Since the word "significant" means different things to different people, there will always be disagreement about where—or whether—to draw species lines in exact lineages.[22]

---

[22] Johanson, D, and Edey, M. Lucy: *The beginnings of humankind.* New York: Simon & Schuster; 1981. p. 144.

Johanson shows that he is aware of the danger of bias:

Some other biases are not so healthy. In everybody who is looking for hominids there is a strong urge to learn more about where the human line started. If you are working back at around three million [years], as I was, that is very seductive, because you begin to get an idea that that is where Homo [referring to the genus of humans] did start. You begin straining your eyes to find Homo traits in fossils of that age.[23]

Johanson seems to be well aware of the "low confidence" nature of his work:

I remember wishing that there was some orderly way that we could proceed, a safe retreat into the step-by-step procedures that all scientists find so useful—and so reassuring; something we could work on as technicians to relieve ourselves of the hugeness and the cloudiness of the ideas that these fossils kept pushing into our minds.[24]

Appendix A discusses the six criteria of high-confidence science in terms of well-established hierarchical levels of scientific evidence. Johanson offers up a very different hierarchy of scientific evidence from the viewpoint of a paleoanthropologist:

When bones are scarce, speculation about them can be as daring as one cares to make it, and no one can contradict the speculator. When bones become more numerous—when a

---

[23] Ibid, p. 257.
[24] Ibid, p. 261.

single fossil is augmented by a large sample of fragments from a number of individuals—those fragments begin to make assertions about themselves that forbid some earlier speculations. The sheer increase of information cuts off a number of possibilities as to what they might or might not be or what they might do. On the other hand, a good assemblage of bones increases the respectability of certain other speculations. On better evidence they improve themselves from merely hopeful guesses to logical probabilities. Once in a great while, a set of bones provides a certainty.[25]

Johanson's book includes an important subtitle: *How Our Oldest Human Ancestor Was Discovered and Who She Was*. This subtitle clearly claims that Lucy is an ancestor to humans—in other words, that she lies in the evolutionary path to humankind. This claim motivates the interest in studying Lucy, but is it grounded in high-confidence science? We would like to apply the tool of science to answer the question, "Is Lucy [or any other fossilized primate] a direct ancestor to humans?" Although several hundred specimens of *Australopithecus afarensis* have been found, the fact that repeated specimens have been found does not address our question. Unfortunately, the hypothetical evolution of humans cannot be repeated to determine if Lucy is a direct ancestor to humans. This meets the 1st criterion of low-confidence science. Obtaining a sample of fossilized bones and observing their similarity to humans is a very indirect way of inferring an evolutionary lineage. We would like to directly measure the evolution of humans from our ancestors, but all that we can do is indirectly measure the similarity of a few bones. This meets the 2nd criterion of low-confidence science. The study of fossilized bones is inherently

---

[25] Ibid, pp. 180–181.

retrospective and observational, which is the 3rd criterion of low-confidence science. A list of similarities between the bones of Lucy and those of modern humans can only suggest an association; it cannot demonstrate a causality such as "this evolved from that."

With such indirect evidence, the influence of bias can easily overwhelm the influence of the evidence (fulfilling the 4th criterion of low-confidence science). The above quotes from Johanson speak to this influence of bias. In Johanson's sequel book, *Lucy's Legacy: The Quest for Human Origins*, Johanson calls himself a "lifelong atheist."[26] This provides insight into his own bias: that of desiring to support his beliefs through his research. Assumptions are essential when interpreting such indirect evidence (fulfilling the 5th criterion of low-confidence science). Although Johanson's books never provide a direct argument that Lucy evolved into humans, the subtitle of his first book makes it clear that this argument is implied. Johanson therefore makes the most fundamental of all assumptions in science: that of assuming that your hypothesis is true. Finally, because no direct ancestral relationship has been demonstrated, the words "our oldest human ancestor" in the subtitle of his first book are an overstated scope of the results. The results that he obtained include a collection of old bones that have similarities to human bones. His interpretation of the results—that Lucy is an ancestor to humans—is clearly overstated, thus fulfilling the 6th and final criterion of low-confidence science. We therefore see that claiming Lucy to be a direct ancestor to humans meets none of the six criteria for high-confidence science.

As of July 2016, the web page for the Smithsonian Museum of Natural History contained the following image of the "Human Family

---

[26] Johanson, D, and Wong, K. Lucy's legacy: The quest for human origins. New York: Three Rivers Press; 2009, p. 61.

Tree" (Figure 6).[27] The top branch of the tree is the "Homo group," including humans. The Homo group contains several different designated species of the genus "Homo"; but who is to say that they could not reproduce with one another?[28] This would, by definition, make them all the same species. Lucy appears on a lower branch of the tree called "Australopithecus group" (she appears in the far right of this group). The most interesting part of this diagram is that Lucy is not depicted as an ancestor to humans; she appears on a side branch, not in the direct line to humans. In fact, *nothing* is depicted in the direct line to humans, except an extremely empty and vague tree branch. In other words, the image makes no claims to any form of ancestral lineage. All fossil evidence other than Homo is depicted as a series of side branches to human evolution. To their credit, the Smithsonian Museum of Natural History admits here that direct ancestry cannot be addressed with any level of scientific confidence. This stands in stark contrast to the message of the Google Doodle and the subtitle of Johanson's book. The fact is, the evidence that Lucy or any other fossilized species lies directly in the path of human evolution is of such low confidence that a consensus cannot be reached by the scientific community.

The format of this diagram will undoubtedly "evolve" over time, just as it has been modified repeatedly over the last century. In addition, many other depictions using different organizations of the branches of life are available. The instability of how this figure is presented, with many debates and new discoveries causing continual rearrangements, speaks to the low confidence of the science.

---

[27]   Human Origins Program, Smithsonian Institution: http://humanorigins.si.edu/evidence/human-family-tree. Reprinted with permission.

[28]   https://www.newscientist.com/article/dn26435-thoroughly-modern-humans-interbred-with-neanderthals/.

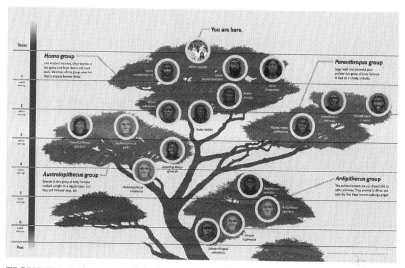

FIGURE 6: A depiction of the human family tree according to the Smithsonian Museum of Natural History, 2016.

VESTIGIAL FEATURES

A common argument in support of macroevolution is the presence of vestigial organs: seemingly purposeless leftovers from prehistoric ancestors. Common examples include the wings of flightless birds and insects, eye remnants of blind fish or moles, pelvis and leg bones in whales, and the appendix in humans. If we are to apply the scientific method to this line of evidence, we would be trying to answer a question such as: "Are vestigial organs a result of macroevolution?"

We first have to decide what is, and what is not, a vestigial organ. How much function is required for an organ to rise above the pejorative label "vestigial?" The more we learn about physiology the more we appreciate the complex function of organs, and the "vestigial" label fades. Obviously, classifying an organ as "vestigial" is subjective and therefore is conducive to demonstrating bias (the 4th criterion of low-confidence science). We also need to keep in mind that the concept of loss of function to create vestigial organs is actually evidence of *devolution*, not evolution. Vestigial organs are thus positioned as a "side effect" of macroevolution and are, at best, very indirect evidence of the process of macroevolution (the 2nd criterion of low-confidence science). The argument for vestigial organs also makes a fundamental assumption (the 5th criterion of low-confidence science) that a creation explanation for life precludes an organ from experiencing a decrease in function over many generations. A belief in creation should not prohibit a decrease in function over time; in fact, the creationist concept of "genetic entropy" suggests that the genomes of all living organisms are degrading over generations.[29] Fish living in caves and losing vision and birds or insects that become flightless therefore do not preferentially support macroevolution over creation. This substantially narrows the applicability of the argument that vestigial organs are evidence of macroevolution: only those cases of vestigial organs that are closely associated with a macroevolutionary process could be relevant. This requires one type of organism to evolve into a very different type of organism, while a function is lost but a vestigial organ remains. Possible examples here include pelvis and leg bones in whales[30] and the appendix in humans.[31] To address this with high-confidence science, you would have to study an organism over thousands of generations and observe, in response to some selective intervention, a process of macroevolution while observing an organ losing function. Of course,

---

[29] Sanford, JC. Genetic entropy. Fourth edition. FMS Publications; 2014.

[30] A recent study suggests that the "vestigial" pelvic bones of whales and dolphins play an important function in reproduction: Dines, JP. Sexual selection targets cetacean pelvic bones. Evolution. 2014;68:3296–3306.

[31] In mammals, the size of the appendix appears to be related to the consumption of cellulose (leaves). Our small appendix thus provides evidence that we're not good at eating leaves; it does not provide high-confidence evidence that we evolved from mammals that have a larger appendix than ours.

if you did this, the direct observation of macroevolution would be much more pertinent than the observation that an organ has lost its function. Unfortunately, direct observation of macroevolution is not possible, so the fossil record is offered as a substitution. Invoking the fossil record leads us back to the above argument for low-confidence science of the fossil record. We must therefore conclude that the question "Are vestigial organs a result of macroevolution?" can only be addressed with very low-confidence science.

Another line of evidence for vestigial remains comes from the field of genetics. When the Human Genome Project was first completed in 2003, initial estimates of the function of human DNA suggested that only 1.2 percent of human DNA had a function; the specific function was coding proteins or regulating the expression of proteins. Those who believe in grand evolution hailed this as strong evidence of macroevolution and coined the term "junk DNA," implying that most of our DNA consists of vestigial remnants of our evolutionary past. In the decade following the completion of the Human Genome Project, the ENCODE study (an acronym for "Encyclopedia of DNA Elements") vastly improved our appreciation of the functional elements of human DNA. In a series of thirty papers published in 2012, the ENCODE study shocked the world by demonstrating that more than 80 percent of the human genome now had known functions (compared to the original 1.2 percent). The >80 percent figure exceeded even the most optimistic estimates, because evolutionary assumptions instilled expectations for the accumulation of junk DNA. The >80 percent proportion of functional DNA will only increase over time as we continue to learn more. The term "junk DNA" is now a bit of an embarrassment, which is rarely mentioned. With the demise of the term "junk DNA," we also witness the demise of the argument that vestigial remains of DNA are evidence of macroevolution.

HOMOLOGY

More than a century before Darwin's time, Carl Linnaeus organized all life-forms. Linnaeus grouped life-forms with similar characteristics together and organized them into levels of structure called "taxonomies." The most familiar levels of taxonomy are called genus and species; a common example is *Homo sapiens*. The organizational structure that Linnaeus proposed is largely maintained to this day. After Darwin, a different interpretation of the taxonomic structure was offered. Organisms with similar traits were considered to be closer neighbors on an evolutionary tree of life. In other words, their traits were similar because they shared a more recent common ancestor. After James Watson and Francis Crick (discoverers of the structure of DNA), appreciation for the genetic similarity of life-forms opened another dimension. Not surprisingly, organisms with similar observable traits (*phenotype*) tend to have similar genetic code (*genotype*). The newly observed similarities in genotype between organisms with similar phenotype was seen to support the evolutionary interpretation of taxonomic structure.

The similarity of either the phenotypes or genotypes of life-forms, often called "homology," is commonly cited as evidence for macroevolution, because such homology is seen to support common ancestry. The argument goes like this: If life evolved from common ancestors, we would expect life-forms to maintain phenotypic and genotypic similarities down the branches of a given limb of the tree, and that is exactly what we see. Richard Dawkins makes a bold claim:

> Just as the vertebrate skeleton is invariant across all vertebrates while the individual bones differ, and just as the crustacean exoskeleton is invariant across all crustaceans while the

individual "tubes" vary, so the DNA code is invariant across all living creatures, while the individual genes themselves vary. This is a truly astounding fact, which shows more clearly than anything else that all living creatures are descended from a single ancestor.[32]

Bill Nye agrees: "This business of homology is one of the absolutely most compelling indicators of the process of evolution."[33]

I think we can all agree that common ancestry is one possible explanation for homology. A second possible explanation is design. Clearly, the vast majority of new products designed by humans reuse components or concepts from other products. As a designer of medical products, I can attest that new designs almost always come from modifications of prior designs, or at least from reusing prior components. A designer of life could possibly take a similar approach, thus producing the homology that we observe. A third possible explanation for homology is convergent evolution. Convergent evolution involves the appearance of similar features in two groups of organisms, yet their common ancestor did not have these features. In other words, the features evolved independently in the two groups, but the features happen to be similar because the process of evolution sculpted them in a very similar, yet independent way. Commonly cited examples include the great similarity between some placental (eutherian) mammals and marsupial mammals, like the wolf and the recently extinct marsupial wolf (thylacine); the mole and the marsupial mole; the groundhog and the wombat; the rabbit and the bandicoot; and the flying squirrel and the sugar glider. Thus, we now have three potential explanations for homology: common ancestry, design, and convergent evolution.

---

[32] Dawkins, R. The greatest show on Earth. New York: Free Press; 2009, p. 315.

[33] Nye, B. Undeniable: Evolution and the science of creation. New York: Saint Martin's Press; 2014, p. 149.

In order for the argument of homology to preferentially support common ancestry over design, a fundamental assumption must be made (the 5th criterion of low-confidence science): a creator or designer would not reuse concepts between organisms. Neither Dawkins nor Nye, nor any biology textbook that uses homology as evidence for common ancestry, recognize this critical assumption. This assumption cannot be tested or validated. As a result, it is impossible for observations of homology to preferentially support common ancestry over design. The argument that homology implies common ancestry therefore degrades to something like: "The evidence of homology does not contradict the concept of common ancestry."

It is important to appreciate, however, that the commonly cited examples of homology are carefully selected to support the argument of common ancestry. There are many striking examples that speak to the lack of homology between life-forms, such as orphan genes (i.e., genes that are unique to a particular species, see chapter 8) and the wide gap between all eukaryotes and all prokaryotes (discussed in chapter 9). Selecting only the evidence that supports common ancestry, and ignoring the evidence against it, is a type of bias—the 4th criterion of low-confidence science. As a result, the argument for homology further degrades to something like: "A carefully selected group of observations—those that show similarity between life-forms—do not contradict the concept of common ancestry."

Furthermore, if homology is to be used as evidence in support of common ancestry, we must rule out convergent evolution. The only way to rule out convergent evolution, however, is to show that the homologous trait was present in the common ancestor. This would require finding the common ancestor. Here the argument of homology runs into a dead end: homology is presented as evidence for common ancestry, because the common ancestors cannot be observed directly. But homology can only support the idea of common ancestry if the common ancestors are observed directly (to show that they contain the homologous traits—thus ruling out convergent evolution).

We now have a different perspective on the above quotes from Dawkins and Nye. They claimed that homology is the clearest and most compelling evidence for common ancestry and macroevolution. Because we have just seen that homology does not preferentially support common ancestry and macroevolution, Dawkins and Nye thus have two choices: 1) admit that their statements are inaccurate or 2) maintain that their statements are accurate, which implies that common ancestry and macroevolution are not well supported by any line of evidence (because the clearest and most compelling evidence does not support their case).

Can observations of homology (either phenotype or genotype) be considered high-confidence evidence of macroevolution? While the argument of the preceding paragraphs largely put this to rest, for the sake of completeness we should attempt to apply the six criteria of high-confidence science to observations of homology. A question like, "Do similarities in phenotype and genotype exist across lineages of life-forms?" could be addressed with high confidence. However, this question does not preferentially support evolution over design. If the desire is to preferentially support evolution over design, the question becomes something like, "Are similarities between life-forms a result of macroevolution?" Here, we are attempting to determine what process is responsible for producing observable life-forms by observing similarities between life-forms. Much like the above arguments for the fossil record, this is an extremely indirect means of determining the process that is responsible for producing life-forms; it is so indirect that it is quite difficult to apply the six criteria.

Criteria of Low-Confidence Science	How they apply to "Are similarities between life-forms a result of macroevolution?"
1. Can't be repeated	The process that produced existing life-forms cannot be repeated.
2. Indirectly measured, extrapolated, or inaccurate results	Morphological similarities between life-forms can be directly measured, but that doesn't address our question. Observable examples of speciation are microevolution, which does not address the question (see chapter 7). To address the question with direct measurement, we would need to observe the process that produced the life-forms.
3. Retrospective, observational study	The duration of a prospective interventional study is prohibitive (millions of years?). Directly observed examples of speciation are microevolution - they do not address the question. The only other option is to observe homology in the fossil record, which is retrospective and observational.
4. Clear opportunities for bias	When homology is cited as evidence for macroevolution, only examples of similarity across life-forms are mentioned. Differences are ignored. Selecting only the evidence that supports a theory is an extreme form of bias.
5. Many assumptions required	To support macroevolution over design by a creator, one must assume that a creator would not reuse concepts between life-forms.
6. Overstated confidence or scope of results	Any attempt to state that homology preferentially supports macroevolution over another process of producing new life-forms is overstated.

Another type of homology that is commonly cited as evidence for macroevolution includes the similarities observed in embryology, the early developmental stages of an organism. The argument is that early embryological stages of very different life-forms have a similar appearance. Indeed, all of life starts out as a single cell, so the resemblance between different life-forms is striking! The actual claim is that fish, reptiles, and humans have similar appearances at certain intermediate stages of their embryonic development. The German biologist and philosopher Ernst Haeckel (1834–1919) is famous for proposing this idea and taking it one step further by proposing that as an embryo develops, it goes through a sequence of stages that reflect the sequence of its evolution. He gave this hypothesis a very impressive-sounding name: "ontogeny recapitulates phylogeny."

We would like to know if the question, "Are similarities in embryonic form a result of macroevolution?" could be addressed by high-confidence science. This follows a parallel argument to the similarities between life-forms. In short, the similarity of embryonic forms is very indirect evidence of macroevolution (the 2nd criterion of low-confidence science). The argument shows bias by emphasizing similarities and dismissing differences (the 4th criterion). Also, the judgment of "similarity" is subjective, which opens the door for bias. Finally, to provide preferential support for macroevolution over design, one must assume (the 5th criterion) that a designer would not have embryos look similar at any stage of development. Embryonic homology can thus only provide low-confidence evidence for macroevolution.

BAD DESIGN

Among the evidence commonly cited for macroevolution are observations that some features of living organisms appear to be poorly designed. Human examples include the appendix, the light-sensing structures of the eye, and the recurrent laryngeal nerve. We've already considered the claims about the human appendix being a vestigial organ. The argument for bad design goes one step further: because the appendix can become clogged and infected, which can lead to death if untreated, it is poorly designed.

The retina of the eye contains specialized cells (rods and cones) that convert light into nerve impulses. One would expect an arrangement in which these cells would be the first to have contact with the light. We might expect the nerve connections to be located behind the light-sensing cells so that the nerves would not get in the way. The human eye, however, is the opposite. The light must first pass through the nerve fibers and several layers of nerve cells (and even through some capillaries) before it reaches the light-sensing cells. The light-sensing cells are also backward, such that the light-sensing part of the cell is at the far end. Although the retinal layers in front of the light-sensing cells are largely transparent, this arrangement would seem to reduce the sensitivity and acuity of the light-sensing cells. The nerve fibers also gather together at a central location to exit the retina. There is no room for light-sensing cells in this location: this is the "blind spot" of the retina. Interestingly, many aquatic invertebrates, for example the octopus, have the more expected retinal arrangement.

Finally, the recurrent laryngeal nerve is a branch of a cranial nerve that controls our voice box (larynx). Because this nerve travels from the head to the larynx, one would expect a direct path. However, the recurrent laryngeal nerve travels down into the thorax, performs a U-turn around the aorta (left side) or the right subclavian artery (right side), and heads back up the neck to the larynx—hence the name "recurrent." This unexpected arrangement seems wasteful, and could slow nerve conduction.

Because evolution does not progress with conscious intention toward a goal, one can imagine that evolution would produce a few results that were suboptimal but "good enough" for survival. This can occur because the process of evolution starts down a path somewhat randomly but then progresses down that path and becomes relatively committed to that approach. Because the retina is so complex, the process of inverting it to improve the arrangement is now very complex and very unlikely to happen. Thus, according to the argument of bad design, the process of macroevolution has left us with many suboptimal features.

Like the argument for homology, this argument relies on assumptions (the 5th criterion of low-confidence science). The argument first assumes that we are smart enough to know what is, and what is not, a bad design. It is incredibly easy to criticize a design by pointing out aspects that we believe are suboptimal. If you walk a mile in the shoes of the designer, however, you might see the method to the madness and agree that a good design trade-off was made.

Let me give you an example. We all start out life with one inbred mechanism for drinking, and the mechanism literally sucks. For most of us, sometime after we pass our first birthday, someone will place a cup of liquid in front of us for the first time. Every baby that I have ever seen will place his or her lips on the rim of the upright cup and start sucking, which of course does not work. Although babies cannot articulate what they are thinking, you can imagine that they are criticizing the bad design of this drinking method. After another year or so, toddlers have learned to appreciate the finer qualities of drinking from a cup, raising the glass up and tilting it to let gravity help in the drinking process. They have fully transitioned to the new and improved method of drinking. But then, someone gives them a cup of liquid with a straw in it. This is a form of double-crossing. Every toddler I have seen try this for the first time will place his or her lips on the straw and then lift the glass up and tilt it, thus dumping the drink on themselves. Once again, they are likely to conclude that this drinking-straw idea is a bad design; only later to find that this design has a good purpose. In addition to having personally endured these humiliating experiences, I can assure you that as an adult I have frequently criticized a design, only to find out later that it is actually quite good. Proof of bad design can only come from creating and validating a better design. Without validation, claims of bad design are inherently subjective and are quite open to bias—the 4th criterion of low-confidence science.

The argument for bad design has a second inherent assumption: a feature that was designed could not have degraded over time to become what we observe (e.g., the appendix). In contrast to this assumption, the designer could have started with a better design, but generations of accumulated mutations could have led to what we see today; this is an example of the creationist concept of "genetic entropy," which, as noted earlier, is the idea that the genomes of all living organisms degrade over generations. Thus, "bad design" may not preferentially support macroevolution over creation.

By citing observations of apparent bad design as evidence for macroevolution, the tool of science is being applied to answer the question, "Are examples of apparent 'bad design' a result of macroevolution?" For completeness, the six criteria of low-confidence science are reviewed in the table below.

BIOGEOGRAPHY

Biogeography is a branch of biology that deals with the geographic distribution of life-forms; it is the science of "what lives where." By exploring the world and observing what life-forms naturally exist in each location, the inquiring mind will seek an explanation for the observations. Some observations are easy to explain: Why are there no penguins in Kansas? Other observations are more puzzling: When humans first explored New Zealand, why did they not find mammals other than a few bats and marine mammals like seals? The nonuniformity of life-form distribution is frequently cited as evidence for evolution.

Criteria of Low-Confidence Science	How they apply to "Are examples of apparent 'bad design' a result of macroevolution?"
1. Can't be repeated	The process that produced existing life-forms with "bad designs" cannot be repeated.
2. Indirectly measured, extrapolated, or inaccurate results	To address the question directly, we would need to observe the process that produced existing life-forms with "bad designs". If we could do that, we would have more direct evidence of macroevolution.
3. Retrospective, observational study	A prospective interventional study can be imagined - following many thousands of generations of organisms, attempting to show that macroevolution occurs and results in "bad designs". Unfortunately, the duration of the study (millions of years?) would be prohibitive. So, we are left with retrospective observational study.
4. Clear opportunities for bias	Deciding what is a "bad design" is a subjective assessment, open to bias.
5. Many assumptions required	Two fundamental assumptions are required: 1) we know what is and what is not a bad design, without having created and validated a better design, and 2) a created feature would not decline in function over time to become a "bad design".
6. Overstated confidence or scope of results	Citing examples of 'bad design' as evidence for macroevolution, without stating the inherent assumptions, is overstated confidence.

Perhaps the most interesting cases involve oceanic islands (i.e., islands that were never connected to continents), such as the Hawaiian or Galapagos Islands. Because oceanic islands were formed by volcanic activity or coral reefs, they started off rather barren. When oceanic islands were first explored, the explorers noted that the life-forms generally did not include amphibians, reptiles, freshwater fish, or mammals (except for the aforementioned bats or aquatic mammals), but they did include birds, plants, and insects. Because the life-forms they found were capable of traveling from neighboring continents to the newly formed islands—whereas the missing life-forms generally could not survive a lengthy oceanic voyage—this was seen as evidence for evolution.

Oceanic islands also provide examples of related species that diversify over time to fill ecological gaps, such as the more than twenty species of Hawaiian honeycreepers or the fourteen species of Darwin's finches on the Galapagos Islands. Although the life-forms on oceanic islands are limited, many related species of the limited kinds of life-forms tend to exist. This speciation is also cited as evidence for evolution. Finally, the life-forms that we find on oceanic islands tend to be similar (but not identical to) species found on the nearest continent. This supports the scenario that life-forms traveled from the mainland to the island, either willingly or by accident, and then took up residence there and fanned out into related species. This, too, is cited as evidence for evolution.

In contrast, continental islands—islands that once were connected to neighboring continents, such as Great Britain—generally contain very similar life-forms to those of the neighboring mainland. This is because the life-forms were distributed across the continental island and adjacent mainland before the island was separated from the mainland, either by rising ocean levels or continental drift.

Summarizing the above, there are three lines of evidence that oceanic island biogeography supports evolution: 1) life-forms are limited to those that can traverse great distances of ocean, 2) the limited life-forms that do exist are found in groups of closely related species, and 3) the limited life-forms closely resemble those of the nearest mainland. In order for these three lines of evidence to preferentially support evolution over creation, two fundamental assumptions (the 5th criterion of low-confidence science) about creation are inherent:

1.) A creator would have placed the life-forms where they were found (i.e., they have not distributed themselves on their own) and would have distributed the life-forms equally over the earth.

2.) The creator does not allow the life-forms to change over time.

Evolutionists like Jerry Coyne commonly express the first assumption: "Why, may it be asked, has the supposed creative force produced bats and no other mammals on remote islands?"[34] Darwin himself expressed both assumptions when he said, "Why should the species which are supposed to have been created in the Galapagos Archipelago, and nowhere else, bear so plainly the stamp of affinity to those created in America?" Dawkins also adopts these assumptions: "Why would an all-powerful creator decide to plant his carefully crafted species on islands and continents in exactly the appropriate pattern to suggest, irresistibly, that they have evolved and dispersed from the site of their evolution?"[35] This first assumption, as espoused by Coyne, Darwin, and Dawkins, is very difficult to justify. Why couldn't a created plant or animal traverse the ocean to end up on a newly formed island just like an evolved ani-

---

[34] Coyne, JA. Why evolution is true. London: Penguin Books; 2009, p. 106.
[35] Dawkins, R. The greatest show on Earth. New York: Free Press; 2009, p. 270.

mal could? When a new island is formed, are created bats from adjacent lands denied immigration? This assumption creates a "straw man" fallacy—attacking a very limited view of creationism so that the evidence appears to favor macroevolution.

The second assumption relates to our discussion in chapter 5. As noted earlier, evolutionists commonly think that accepting evolution is an "all or none" belief. In other words, if you don't accept macroevolution, then you must believe that life-forms cannot change in any way over time. As we discussed in chapter 5, however, even the most conservative creationist groups, like Answers in Genesis, openly agree that speciation occurs and is a result of the genetic variability that is built into creation. Thus, in opposition to this second assumption, one can view the observed changes as being entirely consistent with creation. Evolutionists' thinking that belief in creation implies a belief that no changes can occur is an unnecessary constraint that fuels the debate. If both sides of the debate can agree that a belief in creation allows for small changes over time, even to the point of speciation, this would remove the second assumption and would go a long way toward reconciling the differences between evolutionists and creationists.

In the absence of these two assumptions, the observed biogeography of oceanic islands that speaks to microevolution (including speciation) can be equally explained by evolution or by creation. Birds, insects, and tortoises are free to find their way to new islands and subsequently to diversify—even to the point of speciation—in order to fill ecological niches, as long as birds remain birds and tortoises remain tortoises. As a result, only the subset of biogeographical evidence that preferentially supports macroevolution over creation is of interest. The remaining question is thus: "Is the distribution of life-forms on Earth a result of macroevolution?" The table below shows that this question can only be addressed by low-confidence science.

Criteria of Low-Confidence Science	How they apply to "Is the distribution of life-forms on earth a result of macroevolution?"
1. Can't be repeated	The process that produced existing life-forms cannot be repeated.
2. Indirectly measured, extrapolated, or inaccurate results	The distribution of life-forms on the planet can be directly measured and differences between these life-forms can be directly measured. However, those observations do not address the question of what process produced these life-forms. To address our question, we need to observe the process that produced existing life-forms, which can only be accomplished indirectly through the fossil record.
3. Retrospective, observational study	A prospective interventional study of macroevolution and the resulting distribution of life-forms on a global scale is not feasible. All other approaches will be retrospective and observational.
4. Clear opportunities for bias	Bias is allowed to dominate because the evidence that macroevolution is responsible for the distribution of life-forms is so sparse.
5. Many assumptions required	Only the fossil record can speak to macroevolution. Therefore, the same assumptions that are used to claim that the fossil record supports macroevolution are required here as well.
6. Overstated confidence or scope of results	Claims that biogeographic evidence preferentially supports evolution over creation are overstated.

CHAPTER SUMMARY

The central theme of this chapter is that the commonly cited evidence to support macroevolution does not meet any of the criteria for high-confidence science. In stark contrast to the process of giving a patient a drug and directly measuring a reduction in blood cholesterol—a process that is repeatable, prospective, and directly measurable—the examples reviewed in this chapter attempt to provide preferential support for macroevolution by offering lines of evidence that are very indirect, retrospective, and cannot be repeated. How can the sparse remains of long-dead organisms (fossils) provide proof for the process that produced those life-forms? How can subjective classification of vestigial features possibly offer direct evidence for the process that produced those life-forms? How can carefully selected similarities between life-forms (homology) preferentially support one process of producing life-forms over another process? How can subjective judgments of bad design prove that one type of process for producing life-forms is the reality? How can the observed geographic distribution of life-forms hope to explain the process that produced those life-forms? In each case, the indirect nature of the evidence opens the door for bias and assumptions. In such a highly emotional and contested topic as this, the destructive power of bias and assumptions is amplified.

Although I'm certain that this chapter will raise many objections (see Appendix B for responses to anticipated objections), the chapter is essentially an appeal to take the scientific higher ground when studying evolution. Convincing evidence (i.e., high-confidence evidence) for evolution must come from the type of scientific process that is exemplified by the study of cholesterol reduction. This includes a repeatable experiment conducted prospectively with an intervention, a controlled environment, and directly measurable results. Fortunately, these types of experiments have indeed been conducted.

7

High-Confidence
Evidence for Evolution

As detailed in chapter 6, the commonly cited evidence for macroevolution is very-low-confidence evidence, since it fails on each of the six criteria of high-confidence science. Here we will explore the opposite extreme: high-confidence evidence for evolution.

DOGS

We've already mentioned one area of high-confidence evidence of evolution: dogs. Within a matter of centuries, the artificial selection process of dog breeders has led to hundreds of breeds of dogs, *Canis familiaris*, all arising from the wolf, *Canis lupus*. The variability of features in these breeds is quite impressive, including at least a fiftyfold difference in adult weight between the largest and smallest dogs. This "experiment" of dog breeding has been conducted with high-confidence science: the breeding process is repeatable (mixing breeds produces a repeated outcome), the results are directly measurable, the process of breeding is a prospective interventional study, and the results *can* be summarized with sober judgment, without bias or assumptions. Dogs therefore represent high-confidence evidence of evolution, albeit a very limited scope of evolution (i.e., microevolution).

Presenting the evidence of dog evolution as support of macroevolution immediately degrades the confidence in the science. The aforementioned Jerry Coyne shows us how this is done:

> True, breeders haven't turned a cat into a dog, and laboratory studies haven't turned a bacterium into an amoeba. But, it is foolish to think that these are serious objections to natural selection. Big transformations take time—huge spans of it. To really see the power of selection, we must extrapolate the small changes that selection creates in our lifetime over the millions of years that it has already had to work in nature.[36]

Dawkins offers a similar interpretation (this quote was also used in chapter 5):

> If so much evolutionary change can be achieved in just a few centuries or even decades, just think what might be achieved in ten or a hundred million years.[37]

With these thoughts, Coyne and Dawkins have taken the high-confidence evidence of dog microevolution and have destroyed all six of the criteria of high-confidence science. Perhaps their greatest offense is to the 2nd criterion (directly measurable). They are asking us to extrapolate results that were directly measured (dog breeding over decades or centuries) to imagine something that can't even be measured *indirectly* (ongoing evolution over millions

---

[36] Coyne, JA. Why evolution is true. London: Penguin Books; 2009, p. 24.

[37] Dawkins, R. 2009. *The Greatest Show on Earth.* New York: Free Press p. 37.

of years). Here is a useful analogy: I observe a car pulling away from a red light and measure its velocity increasing from 0 miles per hour to 30 miles per hour in 5 seconds as the car passes over a hill and out of sight. I conclude that the car is accelerating 6 miles per hour each second. I then conclude that after 60 seconds the car will be traveling at 360 miles per hour and that after one hour the car will be traveling at 21,600 miles per hour. Here I am making the assumption that the measured acceleration continues without limits. Coyne and Dawkins are asking you to believe that this very unscientific extrapolation of measured data is indeed the case when we examine dog evolution, and Coyne tells us that it is "foolish to think that these are serious objections." A vivid imagination is required to extrapolate these observations to the realm of macroevolution, and imagination, as we know, has very little to do with high-confidence science.

What about the other five criteria of high-confidence science? Whereas dog breeding is repeatable (the 1st criterion), the process of macroevolution is not. Although dog breeding is a prospective interventional study (the 3rd criterion), Dawkin's and Coyne's references to millions of years points either to the future, which requires imagination, or to evidence from the fossil record, which is retrospective and observational. These quotes come from books that only present evidence in favor of macroevolution and clearly state that their purpose is to convince the reader to believe in macroevolution. These are men on a mission: men of clear bias (the 4th criterion). These quotes imply (but do not disclose) two powerful assumptions (the 5th criterion): 1) despite the known health problems with pedigree dogs, if the breeding process were to continue for millions of years, the animals would somehow be very healthy, and 2) the observed variability in dog characteristics will continue to expand over time, with no limits. These assumptions cannot be tested. Finally, these quotes are the antithesis of sober judgment of results (the 6th criterion) because they claim that any results obtained on a small scale are applicable to a much broader scope. Before Dawkins retired, his job title was "University of Oxford's Professor for Public Understanding of Science". I'd like to invite you to pause here for a moment to appreciate the irony.

Here is a sober scientific interpretation. Although dog breeds are a very impressive demonstration of the extreme variability that can be achieved through artificial selection, they come with a few limitations: First, everyone will agree that all of the breeds of dogs are still dogs; in fact, they are all the same species. As such, the changes that have been directly observed are constrained to the realm of microevolution. Second, the observed changes—especially those that are extreme—are more likely a demonstration of devolution than of evolution. Pedigree dogs are well known to have a large variety of health issues, all of which stem from degradation of the gene pool over generations of breeding.

To obtain high-confidence evidence of evolution on a larger scale, what we really want is a simple organism that reproduces very rapidly—going through thousands of generations and producing vast numbers of organisms—all within one human lifetime. We can then conduct a prospective experiment in which we present the organisms with a challenging intervention (a natural-selection environment) and observe how they change over time. Better than observing changes in behavior or appearance, we need to trace those changes all the way to the source: changes in DNA. To prepare for this, a short and very approachable review of the basic concepts of genetics and molecular biology would be helpful. Those who are familiar with these basic concepts may want to skip the next section.

GENETICS 101

When Darwin published *The Origin of Species* in 1859, he described natural selection as a force that could change organisms over time, but he had no insight into what controlled these changes. Only seven years later Gregor Mendel published his findings on the inheritance of traits, which essentially began the field of genetics. Through a long series of discoveries by preeminent scientists, DNA (deoxyribonucleic acid) was found to be responsible for the inheritance of traits as well as the development and function of all known living organisms. DNA is essentially a library of instructions, coded in molecular form, on how to perform the processes required for life.

James Watson and Francis Crick were the first to describe the molecular structure of DNA.[38] They published their findings in 1953 and were awarded the Nobel Prize in 1962. The DNA molecule has two strands that are bound together to form a shape like a ladder, except that the ladder is twisted to form what is called a "double helix." Each rung of the ladder is called a "base pair," because each rung consists of a pair of components. These components, called nucleotides, come in four types: cytosine (C), guanine (G), adenine (A), or thymine (T). The two components in each rung of the ladder are chemically matched: C is always paired with G, and T is always paired with A. Thus, the two sides of the ladder are matched—if you know one side, you can create the other side. You can think of the nucleotides as letters of an alphabet, such that the sequences of nucleotides in rung after rung of the DNA ladder are like letters of the alphabet that are strung together in sequences to make words.

The complete human genome contains around 3.2 billion base pairs, or 3.2 billion letters that make a very large book of instructions.

---

[38] Watson, JD, and Crick, FH. A structure for deoxyribose nucleic acid. Nature. 1953;171:737–8.

(This book contains only about 300 thousand letters, so you can think of the human genome as 10,000 of these books.) The genome of the malaria parasite *Plasmodium falciparum* (which we will discuss shortly) contains about 23 million base pairs, or about 1 percent of the size of the human genome. The genome of the bacteria *E. coli* (which we will also discuss shortly) contains about 5 million base pairs.

Groups of nucleotides that contain the information to perform a specific function (similar to a paragraph in an instructional book) are called genes. As Mendel discovered, genes are also the functional unit of heredity, meaning that genes tend to be inherited as intact units. Genes are preceded by regions of DNA called "promoters." These promoters contain instructions that control when and how often the gene is used. The function of some genes is to provide instructions for the manufacture of a particular protein. When the promoter says so, the gene will be used to produce a protein through a process that involves transcription and translation. During transcription, the double helix of DNA is "unzipped" to separate the two strands of DNA. The gene on one of the strands is then transcribed (or copied) to make another molecule called ribonucleic acid (RNA). RNA is very similar to DNA but it consists of only a single strand. The transcribed RNA (which now contains a copy of the gene's information) is then transported to a molecular machine for producing proteins.

Proteins are like a chain that is made up of individual links called amino acids. Amino acids are molecules with two parts: one part that is consistent for all amino acids and another part that is unique for each amino acid. When combining a sequence of amino acids to make a protein, the consistent parts bind together to make the chain; the sequence of unique parts are like a variety of pendants hanging off the chain. The sequence of unique parts gives the protein its unique and complex function. Twenty types of amino acids are commonly used to make proteins. The molecular machine that manufactures proteins translates a sequence of nucleotides of RNA into the proper sequence of amino acids for the protein. This translation process is similar to how a computer translates binary numbers into letters of the alphabet. We have noted that there are only four types of nucleotides, but there are twenty different types of amino acids. As such, the translation is not simply one-to-one. A group of two nucleotides could only provide sixteen unique codes, which would be insufficient for describing one of the twenty different amino acids. Each group of three nucleotides thus translates into one amino acid, as shown in the table below.

The molecular machine thus reads each group of three nucleotides, selects the matching amino acid, and adds this to the growing chain that will become a protein. You may have noticed that the code contains a lot of redundancy: for example, CTT, CTC, CTA, CTG, TTA, and TTG all code for the same amino acid (leucine). This was originally thought to be simple redundancy, but a recent paper[39] suggests that the different codes for the same amino acid specify different rates of translation, because the growing protein chain needs time to fold into a three-dimensional shape after the addition of some of the amino acids. Improperly folded proteins are typically useless: similar to the results of any of my attempts at origami. The resulting protein can then serve one of a very wide variety of functions: an enzyme that can facilitate an important chemical reaction, a hormone like insulin that controls metabolism, part of a muscle fiber that contributes to motion, a chemical receptor that receives signals, part of a machine to copy DNA, part of a machine that manufactures proteins, or part of a system that transports molecules according to how they are labelled, just to name a few.

---

[39] Li G-W, et al. The anti-Shine-Dalgarno sequence drives translational pausing and codon choice in bacteria. Nature. 2012:484;538–41.

Amino Acid	Associated Group of Three Nucleotides
isoleucine	ATT, ATC, ATA
Leucine	CTT, CTC, CTA, CTG, TTA, TTG
Valine	GTT, GTC, GTA, GTG
Phenylalanine	TTT, TTC
Methionine	ATG
Cysteine	TGT, TGC
Alanine	GCT, GCC, GCA, GCG
Glycine	GGT, GGC, GGA, GGG
Proline	CCT, CCC, CCA, CCG
Threonine	ACT, ACC, ACA, ACG
Serine	TCT, TCC, TCA, TCG, AGT, AGC
Tyrosine	TAT, TAC
Tryptophan	TGG
Glutamine	CAA, CAG
Asparagine	AAT, AAC
Histidine	CAT, CAC
Glutamate	GAA, GAG
Aspartate	GAT, GAC
Lysine	AAA, AAG
Arginine	CGT, CGC, CGA, CGG, AGA, AGG

The complex function of the protein is highly dependent on having the proper sequence of amino acids, just like the meaning of a word is highly dependent on having the proper sequence of settler (oops, I meant to say "letters"). If the DNA has an error in the sequence of nucleotides, that would be transcribed into an error in the RNA, which could be translated into an error in the amino acid sequence. If we are lucky, this may not have an adverse impact, but it could be devastating: sickle cell anemia results from a single incorrect nucleotide, while cystic fibrosis results from a single missing amino acid. All organisms (including you and me) contain many such errors in our DNA. Some of these errors are inherited directly from our parents, and some of these errors appear in our DNA when the DNA is copied to make an egg or sperm cell. The process of copying DNA is highly accurate, but because the DNA code is so extensive, even rare errors can add up. Approximately 1 out of 100 million nucleotides in a duplicated DNA will have an error. For the human genome, with its 3.2 billion base pairs of DNA, this means that each copy of the entire genome will produce roughly thirty-two errors. Another word for these errors is "mutations."

Mutations come in several different varieties. A "point" mutation is a change in a single nucleotide; a "silent point" mutation changes a single nucleotide, but the translation to an amino acid is unchanged. For example, if a sequence of three nucleotides "GCC" is changed to "GCT," both of these sets of three nucleotides are translated to the amino acid alanine, so the mutation may not matter (unless the translation speed and the protein folding are altered). A "deletion" is a mutation that cuts out a segment of DNA, and an "insertion" is a mutation that adds DNA. If a single base pair is removed or inserted, this "frame shift error" can change all subsequent sets of three nucleotides in the gene, thus resulting in translation to a dramatically different protein product. Finally, a "duplication" mutation results in a repeated segment of DNA.

Returning to how this section began, although Darwin saw natural selection as a force that could change organisms over time, he didn't know what controlled these changes. We now know that DNA is largely responsible for these changes. Mutations in DNA lead to changes in traits, while natural selection serves to filter the traits, selecting those organisms that have favorable combinations of traits for having more offspring.

SCIENTIFIC NOTATION AND LARGE NUMBERS

As we review a series of high-confidence experimental studies of evolution in the pages ahead, we will be discussing very large numbers of organisms. It is sometimes convenient to represent these numbers in scientific (or exponential) notation. Figure 7 is a brief review of scientific notation. Scientific notation is a simplified way of representing very large (or very small) numbers. All of the examples that we will use involve simple powers of 10. The number 10 to the first power is simply 10. The number 100 is the same as 10 times 10, or 10 raised to the second power, represented as 10^2. Looking at the larger numbers in Figure 7, the scientific notation is simply 10 raised to a power, where the power is a count of the number of zeros following the 1 in each number. If a number in scientific notation is multiplied by another number in scientific notation, the power of the resulting product is found by adding the powers of the two multipliers. For example, 10^2 times 10^3 is 10^5 (100 times 1,000 is 100,000).

$10^1 = 10$
$10^2 = 100$
$10^3 = 1,000$ one thousand
$10^4 = 10,000$
$10^5 = 100,000$
$10^6 = 1,000,000$ one million
$10^7 = 10,000,000$
$10^8 = 100,000,000$
$10^9 = 1,000,000,000$ one billion
$10^{10} = 10,000,000,000$
$10^{11} = 100,000,000,000$
$10^{12} = 1,000,000,000,000$ one trillion
$10^{13} = 10,000,000,000,000$
$10^{14} = 100,000,000,000,000$
$10^{15} = 1,000,000,000,000,000$ one quadrillion
$10^{16} = 10,000,000,000,000,000$
$10^{17} = 100,000,000,000,000,000$
$10^{18} = 1,000,000,000,000,000,000$ one quintillion
$10^{19} = 10,000,000,000,000,000,000$
$10^{20} = 100,000,000,000,000,000,000$

FIGURE 7: An explanation of scientific (or exponential) notation.

OBSERVABLE SPECIATION

Despite the great variety of common household dogs, they are all classified as the same genus and species: *Canis familiaris*. In this section we will consider extremely rare occasions in which direct observation of speciation (i.e., the formation of a new species) is possible. As we have previously discussed, both sides of the evolution debate generally agree that speciation has occurred, so speciation is not particularly pertinent to the debate. Nevertheless, it is a form of evolution, and we should review the high-confidence evidence that it occurs.

It is helpful to begin with what defines a species. This is a simple request, but it turns out to be extremely difficult to define exactly when two organisms are the same or different species (recall Jack

Horner's observation that different levels of maturity of the same dinosaur species were originally labeled as different species of dinosaurs). To avoid getting caught up in these issues, we will discuss speciation only for existing organisms that reproduce sexually, because the concept of species is more clearly defined for organisms that reproduce sexually. A species is a group of organisms that can reproduce with one another and produce fertile offspring but are reproductively isolated from other such groups. Because domestic dogs are able to interbreed and produce fertile offspring, they are all the same species. Of course, physical limitations make the breeding of Chihuahuas and Great Danes quite unlikely, but again, we are trying not to get caught up with the definition of species.

Jerry Coyne is an expert in speciation. In his book *Why Evolution Is True*, a chapter titled "The Origin of Species" explains how the process of speciation is very slow and generally cannot be directly measured. The vast majority of examples of speciation are therefore inferred from the fossil record or from biogeography; in short, these are examples of low-confidence evidence. He does mention, however, five known cases of speciation that have occurred during a human lifetime. These examples are a special kind of speciation in which the new species ends up with two complete sets of chromosomes. Of the millions of species of life, only a small fraction are thought to have arisen by this method. One example is the Welsh groundsel—a member of the daisy family. (Some of you may be thinking: Daisies reproduce sexually? Well, yes, they do. We have all seen a poor daisy's petals being plucked one by one to determine if "she loves me" or "she loves me not," but nobody considers the love life of the poor daisy. Shame on us.) This flower carries two sets of chromosomes, one from the common groundsel and the other from the Oxford ragwort. The flower is able to reproduce fertile offspring with other Welsh groundsels but is unable to reproduce with either the common groundsel or the Oxford ragwort. This meets the definition of a separate species and it meets the generalized definition of evolution. The evidence that surrounds the formation of this new species is also obtained with high confidence. In a laboratory environment, combinations of the common groundsel and Oxford ragwort have been shown to produce Welsh groundsels, and the genes and chromosomes of the Welsh groundsel have been shown to be a combination of those from the parent species. This result satisfies the first three criteria of high-confidence science: 1) repeatable; 2) directly measurable; and 3) prospective interventional study. And because it would be difficult to argue that bias or assumptions weighed heavily in such a direct result, criteria four and five are also satisfied. Finally, the result is presented with sober judgment by stating that this specific combination resulted in the formation of the Welsh groundsel, and not by overstating the scope by claiming that this method of speciation applies broadly. Thus, criterion six is met.

We now have our first high-confidence evidence for evolution that includes an understanding of the specific genetic changes. In this case, the genetic changes were a simple combination of the genetics of two other organisms to make a new species.

FRUIT FLIES

The common fruit fly, *Drosophila melanogaster*, has been a mainstay for studies of genetics and evolution for decades. The biologist Molly Burke and her colleagues had the vision (and extreme patience) necessary to conduct a high-confidence study of fruit fly evolution.[40]

---

[40] Burke, MK, et al. Genome-wide analysis of a long-term evolution experiment with *Drosophila*. Nature. 2010:467;587–90.

Their study included 605 generations of fruit flies studied over more than fifteen years (nine to ten days per generation). They studied populations in groups of five, which is a form of repeatability of the experiment (the 1st criterion of high-confidence science). The study was designed prospectively with an intervention (the 3rd criterion): for the five populations of flies in the intervention group, the flies that demonstrated accelerated development and early fertility were selected to produce each successive generation; the flies that developed slowly were excluded. The study authors also included five populations in a control group, which were exposed to the same conditions but without selective breeding. The control group is a way of avoiding bias (the 4th criterion). The outcome consisted of direct measurement (the 2nd criterion) of genetic changes over time compared between the treatment and control groups. After all of this work, the authors concluded (the 6th criterion):

> Despite decades of sustained selection in relatively small, sexually reproducing laboratory populations, selection did not lead to the fixation of newly arising unconditionally advantageous alleles. This is notable because in wild populations we expect the strength of natural selection to be less intense and the environment unlikely to remain constant for 600 generations. Consequently, the probability of fixation in wild populations should be even lower than its likelihood in these experiments.

To paraphrase the results, the study authors found no advantageous genetic changes that were preferentially selected and had thus became predominant in the genome. This result was surprising, because the authors' artificial form of selection in a controlled laboratory was stronger, more consistent, and more focused than the types of selection that occur in the wild. As such, the experimental conditions were a best-case scenario for evolution, but the authors observed nothing significant.

One could argue that 600 generations is not enough. This would only translate to roughly 10,000 years of human evolution. Fortunately, some bacteria reproduce about 100 times as fast as fruit flies, so in a similar expanse of time we could study 60,000 generations in order to find high-confidence evidence of evolution. For this, we have Richard Lenski to thank.

SIXTY-SEVEN THOUSAND GENERATIONS OF *E. COLI*

In 1988, Professor Richard Lenski of Michigan State University started something big. He had the grand vision, patience, and perseverance necessary to design and conduct a prospective interventional study of evolution among bacteria. As we will see, this study was brilliantly designed to produce very-high-confidence results—on par with the quintessential example of cholesterol-reducing drugs that we discussed in chapter 3. Lenski's study remains one of the best, if not *the* best, high-confidence scientific study of evolution that has ever been conducted.

It started with twelve identical populations of bacteria called *Escherichia coli* (*E. coli*). Lenski called these bacteria the "founding fathers," because they were pioneers, set apart in a very controlled environment in which it would be easy to observe changes over time. Lenski split the bacteria into twelve populations to see if they would evolve in similar or divergent directions. The controlled environment consisted of a warm bath of a solution called Davis Minimal Broth[41] with a small amount of added glucose as the source of energy. The glucose and other nutrients were intentionally limited to establish a competition for the bacteria. This limitation of nutrients was a very important part of the experiment. The bacteria that were faster to grow, or the bacteria that could make use of nutrients that others could not, would dominate. This is natural selection—survival of the fittest. Although the bacteria were free to partake of the limited nutrients, their rapid reproduction and growth quickly depleted the available nutrients within a twenty-four-hour period. At that point, after six or seven generations of *E. coli* had been produced, reproduction and growth came to a halt. To keep the experiment going, a small portion of each flask (containing a small fraction of the bacteria from the previous day's growth) was withdrawn and added to a new flask with a fresh food supply; this happened every twenty-four hours. The experiment is now nearing its 10,000th day (almost thirty years!) of continuous operation, with a total of 67,000 generations of *E. coli* studied. On every seventy-fifth day, a portion of the bacteria from each flask is frozen, which is another very important part of the experimental design. These bacteria can be frozen in time and can be revived at will in order to repeat or replay a portion of the experiment. Perhaps you've heard that repeatability is an important part of high-confidence science.

We learned above that each nucleotide of DNA that is copied has about a 1 in 100 million (10^8) chance of a point mutation. Thus, if 100 million organisms are produced, chances are that each nucleotide in the genome will have experienced a point mutation in at least one

---

[41]   Carlton, BC, and Brown, BJ. Gene mutation. In: Gerhardt P, ed. Manual of Methods for General Bacteriology. Washington, DC: American Society for Microbiology; 1981: 222–42.

of the organisms. The total number of *E. coli* organisms that have lived under Lenski's experiment is roughly 6×10^{13} (60 trillion). Thus, it is almost certain that each and every one of the approximately 5 million base pairs of DNA in the *E. coli* has been changed in at least one organism during the experiment. This can be viewed as an exhaustive test to find any potential benefit of any single point mutation.

The results? It is very clear that changes have occurred. According to our generalized definition of evolution—changes in the properties of organisms that occur over more than one lifetime—we can conclude with very high confidence that generalized evolution has occurred. The bacteria adapted to compete in this particular environment. The size of the bacteria increased over time, their growth rate on glucose increased, and their speed of exiting a dormant state and returning to a growth state when introduced to a new flask increased. However, these changes quickly reached new plateaus, which demonstrates the limits to how far *E. coli* can be stretched by this environment. Numerous research papers have been written about these changes.[42] Although the changes that occurred tended to be beneficial in this particular environment, this does not imply that the changes would be beneficial in a more natural setting. Several of the changes are likely to be detrimental to bacteria if they are placed back in a natural environment. For example, four of the twelve strains developed defects in their DNA repair mechanisms,[43] and, after only two thousand generations, all twelve strains lost the ability to manufacture D-ribose, a component of RNA.[44]

---

[42] A list of publications is maintained at: http://myxo.css.msu.edu/PublicationSearchResults.php?group=aad.

[43] Sniegowski, PD, Gerrish, PJ, and Lenski, RE. Evolution of high mutation rates in experimental populations of *Escherichia coli*. Nature. 1997;387:703–5.

[44] Cooper, VS, et al. Mechanisms causing rapid and parallel losses of ribose catabolism in evolving populations of *Escherichia coli*. J Bacteriol. 2001;183:2834–41.

Our focus will be on the most significant change that occurred, because we are interested in knowing how far generalized evolution can go. After about 33,000 generations, one of the daily flasks of bacteria was surprisingly cloudy. This could be either a result of contamination or some important change in the *E. coli*. Lenski's team was able to rule out the presence of contaminants so that they could conclude that the *E. coli* had changed in some significant way. The cloudy appearance of the flask stemmed from a greatly increased population of *E. coli*. Given the limited availability of glucose, which previously constrained population sizes, the *E. coli* must have found another source of energy in the flask. The limited nutrient bath contained one likely new source of energy: citrate. Citrate was abundant in the bath, but it was not intended to be a nutrient—it was included as a "chelating agent," which is a way of providing a protective packaging for metal ions in solution. By extracting some of these *E. coli* and placing them in a flask with only citrate as a nutrient, the study authors were able to prove that the *E. coli* had indeed evolved the ability to metabolize citrate. It is well known among biologists that *E. coli* cannot metabolize citrate when oxygen is present,[45] which is a feature that helps to define them as a species.[46]

In their first report of novel citrate metabolism in *E. coli*,[47] the authors did not know the specific genetic changes that were responsible but determined that the changes must have involved at least two mutational steps. They also realized that the acquisition of this capability was extremely rare—more than 10 trillion (10^{13}) *E. coli* in the earlier generations had failed to develop the ability to metabolize citrate.

[45] Lütgens, M, and Gottschalk, G. Why a co-substrate is required for anaerobic growth of *Escherichia coli* on citrate. J Gen Microbiol. 1980;119:63–70.

[46] Scheutz, F, and Strockbine, NA. In: Garrity GM, et al., eds., Bergey's Manual of Systematic Bacteriology, Volume 2: The Proteobacteria. Springer; 2005: 607–24.

[47] Blount, ZD, et al. Historical contingency and the evolution of a key innovation in an experimental population of *Escherichia coli*. PNAS. 2008;105:7899–906.

The discovery of a newly evolved metabolic path sent ripples through the scientific community. Although the first research paper described the results with sober judgment (one of our six criteria of high-confidence science), a summary by Dawkins greatly amplified the findings (emphasis added; "they" refers to creationists):

Not only does it show evolution in action; not only does it show *new information entering genomes* without the intervention of a designer, which is something they have all been told to deny is possible; not only does it demonstrate the power of natural selection to put together *combinations of new genes* that, by the naive calculations so beloved of creationists, should be tantamount to impossible; it also undermines their central dogma of "irreducible complexity."[48,49]

Dawkins made these bold assertions before the actual genetic changes were determined. His claims that the changes included "new information entering genomes" and "combinations of new genes" were actually hopeful predictions at this point, not established observations.

Over the next seven years the exact genetic mechanisms behind the change were published in a series of two papers.[50,51] First, a little

---

[48] Dawkins, R. The greatest show on Earth. New York: Free Press; 2009, pp. 130–1.

[49] The term "irreducible complexity" was introduced by Michael Behe in his book *Darwin's Black Box*. Behe defines irreducible complexity as a single system composed of several well-matched, interacting parts that contribute to the basic function, wherein the removal of any one of the parts causes the system to effectively cease functioning.

[50] Blount, ZD, et al. Genomic analysis of a key innovation in an experimental *Escherichia coli* population. Nature. 2012;489:513–20.

[51] Quandt, EM, et al. Recursive genomewide recombination and sequencing reveals a key refinement step in the evolution of a metabolic innovation in *Escherichia coli*. PNAS. 2014;111;2217–22.

background is in order. Normal *E. coli* are able to metabolize citrate, but only if oxygen is not present. Specifically, a single step in the process of metabolizing citrate—that of transporting the citrate molecule into the cell—does not happen in the presence of oxygen. This is because the gene that produces the citrate transporter has a promoter region that is inactivated by oxygen. In the absence of oxygen, this gene is turned on, and the cell can actively transport citrate across the cell membrane.

The first of two mutations occurred at approximately generation 31,500 of Lenski's experiment: a gene for citrate transportation was copied to a new location. As mentioned earlier, the original copy was preceded by a promoter region that was disabled by oxygen. The new copy was preceded by a promoter region that was active in the presence of oxygen. Thus, the citrate transporter could now be produced in the presence of oxygen. This mutation on its own allowed the cell a very limited use of citrate in the presence of oxygen, because for every citrate molecule to be used as fuel, one succinate molecule had to be ejected from the cell as a trade.

The second mutation occurred at approximately generation 33,000. This mutation occurred in the promoter region of a second transporter—one that brings succinate into the cell. The mutation resulted in an eleven-fold increase in expression of this second transporter. The two mutations therefore work synergistically to bring citrate into the cell, because the cell can now bring in succinate and use that as payment to bring citrate into the cell. This process can provide energy, as long as there is citrate and succinate in the medium. The article by Erik Quandt and colleagues showed that these two mutations were sufficient on their own to allow the metabolism of citrate in the presence of oxygen.

To summarize what happened, two relatively minor changes in *E. coli* DNA were sufficient to allow for the metabolism of citrate: the first change involved duplicating an existing gene, and the second involved increased activity of a second existing gene. The first mutation offered a slight benefit, which was important in facilitating the acquisition of the second mutation. These changes occurred over 33,000 generations and approximately 10 trillion, or 10^{13}, total organisms. As it turns out, a previously reported case of *E. coli* that could metabolize citrate in the presence of oxygen had been reported in 1982.[52] Interestingly, this case also found that the change resulted from two mutations, although they were different than the two mutations in the Lenski experiment. In light of these findings, we can see that Dawkins's bold claims, which he made before the actual mechanism had been determined ("new information entering genomes" and "combinations of new genes") were a gross and irresponsible exaggeration of reality. He has again clearly exposed himself as a salesman, not a scientist.

Although the ability to metabolize citrate in the presence of oxygen is clearly beneficial for the particular environment of the Lenski experiment, this does not imply that these two mutations are beneficial in a more natural setting. Both mutations represent a loss of control of gene expression, which can lead to the wasteful production of protein transporters when they are not needed.

Returning to our six criteria for high-confidence science, Lenski's study is trying to answer the question: What evolutionary changes occur in *E. coli* over time, given a controlled and constrained environment? The table below summarizes the argument that Lenski's study is based on high-confidence science.

---

[52] Hall, BG. Chromosomal mutation for citrate utilization by *Escherichia coli* K-12. J. Bacteriol. 1982;151:269–73.

Similarly to chapters 3 and 4, with contrasting extremes of high-confidence science (cholesterol reduction) and low-confidence science (the cause of King Tut's death), we now have contrasting extremes in the study of evolution: high-confidence science in Lenski's experiment and low-confidence science in the attempt to use the fossil record to explain what produced life-forms. Let's continue with other examples of high-confidence evidence of evolution.

MALARIA AND CHLOROQUINE

The epic battle between humans and malaria is the source of additional high-confidence evidence of evolution.[53] As we will see, this "experiment" is substantially larger than Lenski's work—so large that it could never be conducted in a laboratory. Unlike the Lenski experiment, this example is not experimental science: it did not involve an intervention in a controlled setting.[54] But, as we will see, this experiment, when combined with the supplemental work to determine the genetic changes, clearly represents high-confidence science.

All life on Earth can be categorized into three domains: Archaea, Bacteria, and Eukarya. Archaea and Bacteria are commonly called prokaryotes. *E. coli* is a member of the Bacteria domain. Malaria and humans are members of the Eukarya domain. Eukaryotes have much larger and very much more complex cell structures and DNA than prokaryotes. Eukaryotic cells have a nucleus and organelles, whereas prokaryotes are relatively disorganized. This example of high-confidence evolution is therefore a study of evolution among eukaryotes, which could be very different than what we just learned about *E. coli*.

---

[53] Credit to Michael Behe for raising awareness of this topic in his book *The Edge of Evolution.*

[54] Technically, this study should be classified as a prospective observational study, which is level 5 (of 8) in the hierarchy of scientific evidence discussed in Appendix A.

Criteria of High-Confidence Science	How They Apply To The Question: What evolutionary changes occur in *E. coli* over time, given a controlled and constrained environment?
1. Repeatable	12 separate strains were maintained to look at repeatability. Every 500 generations, a sample was frozen to allow results to be repeated.
2. Directly measurable and accurate results	The cloudy flask after 33,000 generations was directly observed. The extensive genetic analysis, including replaying the mutational steps, provided direct and accurate measurement of the underlying genetic changes.
3. Prospective, interventional study	The study was conducted prospectively. The intervention was the controlled and constrained environment.
4. Careful to avoid bias	Determination of genetic changes is largely an automated and objective process. Bacteria are not biased - they do what comes naturally. Human intervention was minimized.
5. Careful to avoid assumptions	Direct determination of results (for example, genetic sequencing to determine the mutations that occurred) alleviated the need to make assumptions.
6. Sober judgment of results	The papers by Blount and Quandt are written in an objective and highly credible style. Others could exaggerate results as they summarize the findings through their biases (e.g., Dawkins).

Several different single-celled parasites can cause malaria, the most deadly of which is called *Plasmodium falciparum*. For obvious reasons, from here on I'll refer to this particular parasite as either "the parasite" or "malaria." The parasite is carried by a particular type of female mosquito of the genus *Anopheles*. When the mosquito bites, the parasite is injected into a host. The parasite finds its way to the liver, where it reproduces. The individual parasites then attach to red blood cells, which contain their favorite food source: hemoglobin. Red blood cells are essentially bags full of hemoglobin, a virtual *Plasmodium* paradise. The parasite digests the globulin part of the hemoglobin but cannot digest the heme. (This odd piece of trivia will come in handy later.) The parasites rapidly reproduce as they feast on billions of red blood cells. A severely infected person can have 1 trillion (10^{12}) malaria parasites in his or her body. If another mosquito bites, some of the parasites will enter the mosquito along with the blood and will be transmitted to another host when the mosquito strikes again. The vicious cycle then continues.

About 200 million cases of malaria infection occur worldwide each year. Of these, roughly 600,000 people die annually, which amounts to about 1,600 deaths every day. You may recall that King Tut likely had malaria, so this has been going on for quite a long time.

Quinine, a component of the bark of the cinchona tree, was used as a therapy for malaria starting in the seventeenth century. It was effective but had unpleasant side effects. In 1934, a similar drug, chloroquine, was synthesized by Bayer. Chloroquine was more effective and had fewer side effects than quinine. After years of testing, chloroquine was released for widespread use and was extremely effective against malaria for a decade. The first reports of malarial resistance to chloroquine appeared in the early 1960s, arriving almost simultaneously from South America and Southeast Asia.[55] By the 1980s, the majority of cases of malaria could not be cured by chloroquine. It is clear that a change occurred in the parasite, which fits our generalized definition of evolution. Because malaria has such devastating effects,

---

[55] D'Alessandro, U, and Buttiens, H. History and importance of antimalarial drug resistance. Tropical Medicine and International Health. 2001:6;845–8.

and chloroquine is such a valuable drug, scientists have carefully studied the acquisition of chloroquine resistance by malaria.

While Lenski's experiment required about ten trillion, or 10^{13}, total organisms to develop the ability to metabolize citrate in the presence of oxygen, it is estimated that malaria required one hundred million times one trillion, or 10^{20}, total organisms to develop resistance to chloroquine.[56] This number comes from an estimate of 100 million (10^8) humans infected by malaria and treated by chloroquine, multiplied by one trillion (10^{12}) malaria parasites in each human, in order for malaria to develop resistance to chloroquine (see Figure 8). Thus, the malaria "experiment" is 10 million times larger than Lenski's experiment. This should, I hope, give you a feel for the epic scope of this prospective observational study. The evolution of malaria to overcome the selective power of chloroquine must be a very special evolutionary event because of the sheer number of organisms required to accomplish the task.

What genetic changes were required to develop resistance to chloroquine? A series of research reports in the 1990s and 2000s specified the genetic changes (within a number of different strains of malaria) that led to chloroquine resistance.

A little background information is necessary. The malarial parasites ingest the hemoglobin into the "stomach" of the malarial cells. This isn't literally a stomach, but it is an organelle inside the cell that acts like a stomach by breaking down ingested food with enzymes. The globulin parts of hemoglobin are digested, but the heme part cannot be digested. The heme is toxic and will kill the malaria unless the malaria neutralizes it by converting it into crystals. The chloroquine diffuses into the malaria and concentrates inside the stomach. The acidic environment found there essentially traps large quantities of chloroquine in the stomach, which is thought to inhibit the

---

[56] White, NJ. Antimalarial drug resistance. J Clin Invest. 2004;113:1084–92.

neutralization of the heme. Thus, with high concentrations of chloroquine, the undigested heme is toxic to the malaria. One protein that allows the transport of molecules between the stomach and the rest of the cell is the focus of chloroquine resistance. This transport protein evolved in malaria that are resistant to chloroquine so that chloroquine is no longer concentrated inside the stomach, and can no longer interfere with heme neutralization. The transport protein is called "*Plasmodium falciparum* chloroquine resistance transporter."[57] You can probably understand why the abbreviation PfCRT is preferred. The PfCRT protein is a string of 424 amino acids.

10^1 = 10
10^2 = 100
10^3 = 1,000   one thousand
10^4 = 10,000
10^5 = 100,000
10^6 = 1,000,000   one million
10^7 = 10,000,000
10^8 = 100,000,000   ← **Number of humans with malaria, treated by chloroquine, to develop resistance**
10^9 = 1,000,000,000   one billion
10^{10} = 10,000,000,000
10^{11} = 100,000,000,000
10^{12} = 1,000,000,000,000   one trillion ← **Number of malaria organisms in a severely infected human**
10^{13} = 10,000,000,000,000
10^{14} = 100,000,000,000,000
10^{15} = 1,000,000,000,000,000   one quadrillion
10^{16} = 10,000,000,000,000,000
10^{17} = 100,000,000,000,000,000
10^{18} = 1,000,000,000,000,000,000   one quintillion
10^{19} = 10,000,000,000,000,000,000
10^{20} = 100,000,000,000,000,000,000 ← **Number of malaria organisms needed to develop resistance to chloroquine**

FIGURE 8: The number of malaria organisms required to develop resistance to the antibiotic chloroquine.

---

[57] Ecker, A. PfCRT and its role in antimalarial drug resistance. Trends in Parasitology. 2012;28:504–14.

Earlier studies of chloroquine-resistant malaria found consistent substitutions of amino acids 76 and 220 of PfCRT (as a result of point mutations in DNA), accompanied by substitutions in two to six other locations.[58] The consistency of changes in amino acids 76 and 220, from strains around the globe that independently developed resistance, is a type of repeatability of the experiment. The mutation at site 76 was shown to be essential in these cases, because strains that had undergone other mutations but no change at location 76 did not have resistance to chloroquine. We can conclude that four point mutations, and sometimes up to eight point mutations, are sufficient for malaria to develop resistance to chloroquine, and that it took approximately 10^{20} total organisms to accomplish this task. Although this series of mutations is clearly beneficial in the presence of the selective agent chloroquine, the mutations are not beneficial in the wild.[59] If chloroquine-resistant malaria goes for a time without exposure to chloroquine the resistance disappears, because the chloroquine-resistant strains cannot compete with normal strains that are not resistant to chloroquine.

Although malaria's development of resistance to chloroquine is a prospective observational study (not a prospective interventional study), it has all of the other criteria of high-confidence science. The study is repeatable (1st criterion), because chloroquine resistance appeared independently in different geographic locations, with similar genetic changes noted in the resistant strains. The genetic changes associated with chloroquine resistance were directly measurable (2nd criterion). The study was conducted prospectively (3rd criterion), using the administration of chloroquine as an intervention. As

[58] Fidock, DA, et al. Mutations in the *P. falciparum* digestive vacuole transmembrane protein PfCRT and evidence for their role in chloroquine resistance. Molecular Cell. 2000;6:861–71.

[59] Kublin, JG, et al. Reemergence of chloroquine-sensitive *Plasmodium falciparum* malaria after cessation of chloroquine use in Malawi. J Infect Dis. 2003;187:1870–5.

mentioned, however, the study is observational (not interventional), meaning that the study did not control who did and who did not receive treatment. We have discussed the inability of observational studies to demonstrate cause-and-effect. In this case, however, follow-up studies of forced genetic manipulations in the laboratory made it clear that these specific mutations cause chloroquine resistance.[60] The results include genetic sequencing by several different labs, therefore ensuring that they are objective and accurate. The collection of chloroquine-resistant malaria and the comparison of DNA to that of chloroquine-sensitive malaria are unlikely to involve meaningful bias or assumptions (4th and 5th criteria). Finally, the results are presented with sober judgment and proper scope, as they apply only to *Plasmodium falciparum* and to chloroquine resistance (6th criterion).

To add a little more information to this story, more recent chloroquine-resistant strains from Madagascar involve a different protein (not PfCRT),[61] thus showing that there is more than one way to accomplish resistance to chloroquine. The specific mutations in this pathway are not yet known. Malaria has also been able to overcome other antibiotic agents. Approximately one trillion (10^{12}) malarial parasites are needed in order to have one that is able to overcome the antibiotic atovaquone.[62] Similarly, about one trillion (10^{12}) malarial parasites are needed to produce one parasite that is able to overcome

---

[60] Sidhu, AB, et al. Chloroquine resistance in *Plasmodium falciparum* malaria parasites conferred by pfcrt mutations. Science. 2002;298:210–3.

[61] Andriantsoanirina, V. Chloroquine clinical failures in *P. falciparum* malaria are associated with mutant Pfmdr-1, not Pfcrt in Madagascar. PLoS ONE. 2010;5;e13281.

[62] White, NJ. Delaying antimalarial resistance with combination chemotherapy. Parassitologica. 1999;41:301–8.

the antibiotic pyrimethamine.[63] Interestingly, in each of these cases, resistance to the antibiotic is conferred by a single point mutation in the right location.

The table below summarizes what we have learned so far regarding the number of organisms needed to accomplish a particular degree of evolutionary change. Note that the number of required organisms increases dramatically as the degree of evolutionary change increases.

Organism	Selective force	Approximate number of organisms required to evolve	Evolutionary changes
Malaria	atovaquone	1 trillion	1 point mutation
Malaria	pyrimethamine	1 trillion	1 point mutation
E. coli	minimal glucose; abundant citrate	10 trillion	1 gene copied and 1 promoter modified
Malaria	cholorquine	100 million times 1 trillion	4 to 8 point mutations

The high-confidence evidence for evolution that we have reviewed thus far included organisms artificially selected for particular traits (dogs and fruit flies), organisms competing against their peers for limited resources (*E. coli*), and organisms facing a powerful destructive force (malaria). These are different types of selective forces,

---

[63] Wu, Y, et al. Transformation of *Plasmodium falciparum* malaria parasites by homologous integration of plasmids that confer resistance to pyrimethamine. Proc Natl Acad Sci. 1996;93:1130–4.

with different degrees of selective pressure. In each case changes occurred over many generations, thereby satisfying the generalized definition of evolution. However, the known genetic changes in response to these selective pressures had minimal complexity, even after very large numbers of organisms had been produced. We have not seen any completely new genes or proteins. In the largest study (on malaria), the changes were clearly not beneficial in the absence of the selective agent: in the absence of chloroquine, the trait of chloroquine resistance is lost over time. Therefore, although these studies clearly demonstrated changes occurring over generations, they are not very satisfying demonstrations of meaningful improvements in organisms over time: a hallmark of macroevolution.

TRYPTOPHAN PRODUCTION

The next example of high-confidence evidence for evolution is an intentional test of the power of evolution to complete the development of an enzyme (a complex protein that facilitates a chemical reaction). The simplest enzymes contain about 60 amino acids, so about 180 nucleotides of DNA are required to code the simplest enzymes. The immense scope of the malaria study resulted in changes to between 4 and 8 nucleotides—clearly a long way from 180 nucleotides. Thus, the scope of experimentation that can be conducted in a controlled laboratory environment could not produce anything as complex as a new enzyme. But perhaps it would be possible in a laboratory environment to show evolution making a few progressive steps in the development of an enzyme. We'd like to address the question: Can a nearly functional gene be driven to completion by the process of evolution? This question could be addressed by a high-confidence prospective interventional study (the 3rd criterion of high-confidence science). The intervention is a combination of taking an existing gene, inserting mutations to decrease its functionality, and placing the mutant cells in a selective environment that provides an incentive for repairing the gene. Taking an existing gene and inserting mutations to decrease functionality is like giving someone a beautiful new sports car that is fully functional except that the gas pedal and brake pedal are missing. If I received this as a gift, even though I'm not mechanically inclined, I would find a way to fix these flaws so that I would end up with a fully functional sports car. In a similar manner, one would expect the cell, by random mutations and natural selection over many generations, to repair this broken gene. The molecular geneticist Ann Gauger and colleagues performed such a study,[64] using a very similar (high-confidence) approach as that used by Richard Lenski. Before we get into the experiment, a little background would be helpful.

For organisms to grow and reproduce, they need a full assortment of twenty amino acids. If they find themselves in an environment where one or more of the amino acids are not available, the organism could die or may have to lay dormant until the amino acid becomes available. Fortunately, most organisms have the ability to convert one type of amino acid into another. If there is not enough of the amino acid tryptophan to go around, for example, many organisms can produce an enzyme called "tryptophan synthase" to convert the amino acid serine into tryptophan. *E. coli* happens to have this enzyme, but humans do not.[65] In an environment where tryptophan is not available but serine is, the *E. coli* can switch into a mode of producing its own tryptophan from serine to continue happily along its meta-

---

[64] Gauger, AK, et al. Reductive evolution can prevent populations from taking simple adaptive paths to high fitness. BIO-Complexity. 2010;2:1–9.

[65] If humans need tryptophan, they have to eat it. Turkey contains tryptophan, as does chicken, fish, and milk, but the tryptophan in turkey is commonly blamed for making people sleepy after Thanksgiving dinner. However, it is more likely the quantity of food than the quantity of tryptophan that makes one tired after Thanksgiving dinner.

bolic way. Like most enzymes, tryptophan synthase is not a single protein molecule: it is a combination of several molecules (four in this case) that assemble together to make a functional molecular machine. Tryptophan synthase is made of two copies of one protein (the alpha subunits) and two copies of another (the beta subunits).

Gauger's experiment began by modifying the gene for producing the alpha subunit of tryptophan synthase in *E. coli*. They created one modified organism by changing only one nucleotide in the original gene, such that the modified gene produced a protein with one incorrect amino acid at location 60 in the chain of 268 amino acids.[66] We will refer to this organism as "ModA." ModA resulted in a tryptophan synthase enzyme with reduced functionality. Think of this as a sports car with only the brake pedal missing. It can be driven, but it will have reduced functionality. ModA could grow in an environment without tryptophan, but could not compete with the growth rate of normal *E. coli*. The study authors then created a second modified organism ("ModB") by changing a different single nucleotide in the original gene.[67] ModB could not grow without tryptophan: this single nucleotide substitution completely inhibited the function of tryptophan synthase. Think of this as a sports car with only the gas pedal missing. It is not going anywhere.

When either ModA or ModB was placed in an environment with limited tryptophan (an environment where the ability to produce tryptophan would be favored), both were able to repair the one error in the gene to restore full functionality of the tryptophan synthase. In both cases, approximately 100 million organisms were required to spontaneously restore the function of tryptophan synthase.

---

[66] More specifically, this modification is "D60N," because the aspartic acid (symbolized by "D") at location 60 was changed to the amino acid asparagine (symbol N).

[67] More specifically, this modification is "E49V," because the glutamic acid (symbol E) at location 49 was changed to valine (symbol V).

The authors then created a third modified organism ("ModAB") that included both ModA and ModB changes to the original gene. ModAB still had 99.7 percent correct DNA code for the functional gene: like a beautiful new sports car with both the gas pedal and brake pedal missing. Given this, growing ModAB in a tryptophan-limited environment would be expected to first result in repair of the "B" error to produce ModA. Recall that ModA has a slightly functional enzyme, like a sports car with a gas pedal but without a brake pedal. The resulting ModA would then be expected to grow more rapidly and become dominant over ModAB. To prove this, the investigators placed 1 percent ModA together with 99 percent ModAB in a tryptophan-limited environment. After thirty-three generations, ModA had expanded from 1 percent of the population to 23 percent of the population. After ModA became common in the population, the second repair would be expected to follow, resulting in a fully functional enzyme and a very dominant "evolved" strain (like a fully functional sports car). Because both ModA and ModB were each able to evolve back to full functionality, this expected pathway for the step-by-step correction of the two errors in ModAB seemed quite reasonable.

Similarly to the Lenski experiment, Gauger and colleagues started with multiple separate lines of ModAB (fourteen flasks of "founding fathers" in this case) in a nutrient solution with limited tryptophan. Each day, a new flask containing only tryptophan-limited nutrient solution was infected with a small amount of what was left over from the previous day. This yielded between six and seven new generations per day; a sample was frozen every 500 generations to allow the study to be repeated (the 1st criterion of high-confidence science). The study produced three major surprises.

1. After more than 9,000 generations and more than 1 trillion total organisms had been grown, none of the lines of ModAB reverted to a fully functional enzyme.
2. ModA, the result of the first expected evolutionary repair of ModAB, never became common in the gene pool. A few organisms made this slightly beneficial change, but they never grew in popularity, apparently because they were not able to compete with the others in the solution.
3. ModAB evolved in a different direction, which resulted in a doubling of its growth rate in the tryptophan-limited solution after only 500 generations. Both the "A" and "B" errors remained.

The authors then directly measured (2nd criterion) the DNA changes that had occurred. One of the lines of ModAB had completely deleted both the alpha and beta tryptophan synthase genes. Giving up on tryptophan synthase production saved these cells a lot of wasted energy and wasted tryptophan. (Ironically, it takes tryptophan to make the enzyme tryptophan synthase.) Giving up on tryptophan production seemed to be preferable to fixing the two errors in order to achieve tryptophan production. Referring back to my sports car metaphor, because of the high cost of insuring my new useless sports car, I might choose to drop it off at the junkyard rather than bothering to fix the gas and brake pedals. Twelve other lines of ModAB had substantially reduced expression of the tryptophan synthase genes—again, saving energy, but inhibiting progress toward fixing the gene. The final line also did not progress through the expected evolutionary route and could not produce tryptophan. The specific changes that had occurred were not determined, however, because the recombinant

DNA methods that the authors had used to test for expression did not work on the rearranged gene.

Knowing that the first expected evolutionary step—that of repairing the "B" modification to turn ModAB into ModA—resulted in some degree of tryptophan synthase functionality, the investigators wondered why this pathway was not preferred. They then measured the time it took to double the population size under tryptophan-limited conditions. ModAB took 1.6 hours to double its population; ModA was able to double its population in 1.5 hours, which was a clear improvement. For the ModAB strains that had reduced their expression of tryptophan synthase, however, the improvement was even more notable: rather than fixing the "B" error, they were able to double their population in only 1.2 hours. Thus, the benefits of evolving from ModAB to ModA were clearly overpowered by simply reducing the expression of tryptophan synthase in ModAB.

Figure 9 is a diagrammatic summary of the findings. The x-axis shows the progression of time of the evolution experiment, and the y-axis shows the increasing growth rate of an evolved organism; in other words, moving up the y-axis means that populations can double more quickly. Starting with ModAB in the bottom left corner of Figure 9, and given that repairing the first error would increase the growth rate, the *E. coli* were expected to follow an evolutionary path from ModAB to ModA and then from ModA to full functionality. The two steps in this path each conferred a fitness benefit. However, the *E. coli* never succeeded in following this path, because unexpected shortcuts occurred. The shortcuts increased growth rate, but either deleted the gene entirely, which brings the evolution process of this gene to a dramatic STOP, or dramatically reduced its expression, which will interfere with achieving the goal (WRONG WAY).

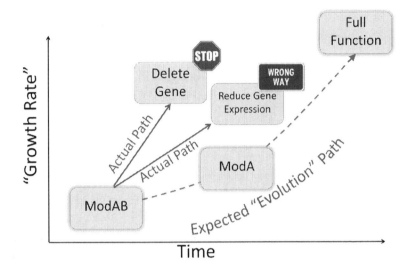

FIGURE 9: *E. coli* with two errors in a tryptophan synthase gene (ModAB) were expected to evolve to fix one error (resulting in "ModA," with improved growth rate in a tryptophan-limited environment) and then fix the other error to arrive at "Full Function." Instead, the *E. coli* took shortcuts to delete the gene entirely or to reduce expression.

We cannot infer from this example that all such evolutionary paths would be sidetracked. This example does show, however, that despite a simple, clear pathway to improved fitness, where each single mutation step along the path confers a benefit, shortcuts that go in the wrong directions can be favored. Given that the alpha subunit of the tryptophan synthase gene is 268 amino acids (804 base pairs), and given that the beta subunit is also necessary for producing tryptophan (the beta subunit is 397 amino acids or 1,191 base pairs), it is sobering to try to imagine a pathway where each single mutation along the path to the functional enzyme could confer a fitness benefit (even if portions are "borrowed" from existing genes) and where no energy-saving sidetrack leads the evolutionary process astray. Certainly, the high-confidence evidence presented in this chapter offers little hope, even for developing a gene that is already 99.7 percent complete.

CHAPTER SUMMARY

We reviewed several high-confidence prospective studies of evolution in this chapter. With the exception of the fruit fly study, each demonstrated changes that occurred over generations. Dog breeding is perhaps the most familiar example, since it exhibits the development of a wide variety of dog breeds in a matter of decades or a few centuries. Those who are eager to support the grand definition of evolution plead for the use of imagination to extrapolate these findings over millions of years. However, although the development of dog breeds is high-confidence science, extrapolation of the observable results instantly negates all six criteria of high-confidence science. Although the great variety of domestic dogs consists of one species, a handful of high-confidence examples of the process of speciation may be found. The examples we have reviewed involved combinations of genetic material from parent organisms, not the development of new genetic components.

Richard Lenski's ongoing study of *E. coli* has documented many minor genetic changes over 67,000 generations. One particular change—the development of an ability to metabolize citrate—was hailed by some as a marvel of evolutionary accomplishment. Upon determination of the genetic changes, however, all of the genetic components were already present; the new ability had resulted from changes in the expression of two existing genes.

In a study that was 10 million times as large as the Lenski study, malarial parasites developed resistance to chloroquine. The genetic changes in this case included an accumulation of four to eight point mutations in one gene. The malaria that evolved resistance were also found to be at a disadvantage when chloroquine was not present.

Finally, an intentional test of evolution to repair a slightly broken tryptophan synthase gene highlighted the limitations of evolution. Although single point mutations could be repaired, a combination of two point mutations proved to be too difficult. Rather than repairing two mutations, the *E. coli* took a lower energy path of reduced expression of the gene.

The measurable genetic changes in these high-confidence studies include combinations of genetic material from two organisms, gene duplications, the enhancement or inhibition of gene expression by mutations in promoters, and a handful of point mutations. These studies therefore provide convincing evidence for the generalized definition of evolution (i.e., changes in the properties of organisms that occur over more than one lifetime). But the studies also provide a clear indication that the process of evolution faces significant constraints. Evolution appears to require very large numbers of organisms and generations, combined with a strong and consistent selective force, in order to achieve relatively minor adjustments of existing genetic components. All of this high-confidence evidence speaks to microevolution, not macroevolution. No completely new genes or proteins were observed. Evolution also appears to take the path of least resistance, which may be sufficient to enhance fitness against a specific selective force but is likely to lead to reduced fitness in the absence of the selective force. These observations, obtained from high-confidence science, do not encourage the adoption of macroevolution—a process that requires significant quantities of new genetic components.

8

Applying High-Confidence Evidence for Evolution

The last two chapters reviewed the high- and low-confidence evidence for evolution. The six criteria for high-confidence science are not intended to produce only black and white designations of high and low confidence. Rather, they are intended to discriminate gradations of confidence levels. However, we found that the evidence for evolution is largely dichotomous. Some forms of evidence provide very low confidence, since they fail each of the six criteria. The remaining forms of evidence provide very-high-confidence, since they fulfill each of the six criteria and match the best approaches that modern evidence-based medicine has to offer (e.g., the study of medications to reduce cholesterol).

This is a turning point in the book. Now that we have completed a review of the evidence and its associated levels of confidence, we can apply the results to assess the validity of macroevolution. When combining evidence from different sources, it is only logical that higher-confidence evidence should be given clear priority over lower-confidence evidence. If I wanted to obtain approval to sell a new drug in the United States and I went to the FDA with results from multiple randomized controlled clinical studies (i.e., very-high-confidence science) and multiple retrospective observational studies (i.e., very-low-confidence science), the FDA would likely ignore the retrospective observational studies. At best, the FDA would consider the retrospective observational trials to be "hypothesis generating," but since the randomized controlled trials had directly addressed the hypothesis with high confidence, the retrospective observational trials would no longer have any value. In a similar manner, we will focus our attention on the high-confidence evidence.

THE SCOPE OF HIGH-CONFIDENCE STUDIES

The high-confidence studies of evolution make it clear that changes do occur in organisms over many generations. In each case, however, the complexity of the evolutionary advancements were limited to a few mutations in a given direction. No completely new genes or new proteins were observed. Despite Dawkins's claims, Lenski's experiment offers no evidence of "combinations of new genes" or anything that approaches the complexity required for a macroevolutionary change.

It is possible, however, that the experiments did not provide enough time to evolve real changes, such as new genes or new proteins. The studies were conducted over decades, whereas macroevolution takes millions of years. If the decades of high-confidence studies were to continue for millions of years, perhaps changes of millionfold complexity should be expected, including completely new genes or proteins.

A first response to this thinking was presented in chapter 7, in the section on dogs. This line of thinking is an appeal to extrapolate available data over great spans of time, which entirely degrades the high confidence of the original studies. Extrapolation like this isn't science; it is wishful thinking. A second response to this line of thinking requires careful consideration of the actual scope of the high-confidence studies. Are they really of such a limited scope? These studies are accelerated prospective demonstrations of evolution, because they involved organisms that reproduce very rapidly and in very large numbers, and because the selective forces were consistent and focused. Let's compare the scope of the high-confidence studies to that of a familiar claim of macroevolution: the evolution of humans from our last common ancestor with chimpanzees. This comparison is not intended to be highly quantitative, because it cannot be. Rather, it is intended to provide a rough approximation of the relative scopes of evolutionary opportunity as a way of putting the high-confidence studies in perspective.

Here is a brief summary of the scope of the high-confidence evolution studies.

- Lenski's *E. coli* experiment included 33,000 generations and more than 10 trillion (10^{13}) organisms to demonstrate evolution of the ability to metabolize citrate in the presence of oxygen (a combination of two mutations).
- The malaria "experiment" included one hundred million times one trillion (10^{20}) total organisms to develop resistance to chloroquine (a combination of four to eight point mutations). The life cycle of *Plasmodium falciparum* is quite complex, so it is not easy to estimate a number of generations in the decades of this experiment. Certainly there were at least thousands of generations.
- Gauger's *E. coli* experiment on tryptophan synthase included 9,300 generations and at least 1 trillion (10^{12}) organisms in an effort to repair two mutations. The desired repair of two mutations was never achieved.
- All studies had a focused, consistent, intense selective pressure as a way of ensuring that natural selection was at work.

For comparison, let's consider the expected scope of the macro-evolutionary process that had to occur to produce humans from our last common ancestor with chimpanzees. This is commonly believed to involve six million years of evolution—clearly much more than the decades involved in the high-confidence studies. However, whereas the *E. coli* experiments produced six to seven generations each day, humans (or hominids) are dramatically slower. The scope of human evolution from our last common ancestor with chimpanzees is expected to involve

- approximately 500,000 generations (assuming twelve years per generation for 6 million years; modern chimpanzees typically produce their first offspring at around thirteen to fourteen years of age);
- approximately 1 trillion (10^{12}) total organisms;[68]
- selective pressures that varied over time and were very likely less intense than those of the high-confidence evolution studies.

---

[68] This number is admittedly a rough guess with a wide margin of error, but the argument here is not very sensitive to the actual number. Here's how the number was obtained: the world population of humans is now (as of 2016) slightly over 7 billion, but historical populations were substantially lower than this. As little as 200 years ago, world population was only 1 billion; estimates of world population diminish rapidly farther back in time. Human population is best represented by an exponentially increasing function, which implies that ancient populations were very small. If you Google "How many humans ever existed," you will see a surprising consensus near 100 billion, which includes the last 50,000 years. To estimate the number of hominids and *Homo sapiens* that existed in the preceding 6 million years, 2 million organisms produced every ten years would lead to a rounded estimate of 1 trillion organisms. This could easily be off by a factor of ten or even one hundred, but, as you will see, the same overall conclusions will be reached.

A comparison of the number of organisms in each of these scenarios is presented in Figure 10. The scope of evolutionary opportunity is a function of

1.) *the total number of organisms.* Each organism that is produced could be endowed with a beneficial mutation: a lucky mistake that could be combined with other mutations to provide a new function. In this sense, each organism is like a lottery ticket: the more organisms that are produced, the greater the chances of winning.

2.) *the number of generations.* Each generation that is exposed to selective pressure(s) is an opportunity for a beneficial mutation to grow in prevalence and achieve fixation or dominance in the gene pool.

3.) *the strength and consistency of selective pressures.* Consistent and more intense selective pressures should accelerate the evolutionary process over inconsistent and milder selective pressures.

Comparing the numbers between the high-confidence evolution studies and the hypothesized evolution of humans, we see that the total number of organisms in the high-confidence evolution studies equals or exceeds the total number of humans/hominids ever to have existed. The studies by Lenski and Gauger had similar total numbers of organisms as the total number of humans/hominids that ever existed, while the malaria experiment included about 100 million times the number of humans/hominids that ever existed. Indeed, each human who is severely infected by malaria can carry as many malaria organisms as the number of humans/hominids that have ever existed!

10^1 = 10
10^2 = 100
10^3 = 1,000 one thousand
10^4 = 10,000
10^5 = 100,000
10^6 = 1,000,000 one million
10^7 = 10,000,000
10^8 = 100,000,000
10^9 = 1,000,000,000 one billion
10^{10} = 10,000,000,000
10^{11} = 100,000,000,000 ♪ **Approx. number of hominids ever to exist.**
10^{12} = 1,000,000,000,000 one trillion ← **Number of E.Coli in tryptophan expt.**
10^{13} = 10,000,000,000,000 ← **Number of Lenski's E. coli needed to**
10^{14} = 100,000,000,000,000 **metabolize citrate**
10^{15} = 1,000,000,000,000,000 one quadrillion
10^{16} = 10,000,000,000,000,000
10^{17} = 100,000,000,000,000,000
10^{18} = 1,000,000,000,000,000,000 one quintillion
10^{19} = 10,000,000,000,000,000,000
10^{20} = 100,000,000,000,000,000,000 ← **Number of malaria needed to develop resistance to chloroquine**

FIGURE 10: A comparison of population sizes: the approximate number of hominids ever to exist is less than or equal to the number of *E. coli* or malaria included in the high-confidence studies of evolution.

In terms of generations, Lenski's experiment had 1/15th of the number of generations of humans/hominids, while Gauger's experiment had about 1/60th of the number of generations of humans/hominids. The complex life cycle of malaria makes it difficult to estimate the number of generations involved.

In terms of selective pressure, the consistent and intense selective pressures of these high-confidence evolution studies should clearly accelerate the rate of evolution relative to human evolution. To put this into perspective, only one malaria organism out of 10^{20} had what was required to survive an attack of chloroquine. In human terms, our world population of 7 billion would be completely wiped out by such a threat. Not a single living human would remain if we faced such a strong selective pressure. In fact, if there were 10 billion earths, each with 10 billion human inhabitants, only one single human being would have what it takes to survive such a threat. Now that is selective pressure! It is very safe to say that the putative evolution of humans from our last common ancestor never faced such a selective force; otherwise, we would not be here today.

In summary, the high-confidence studies of evolution are at least similar in the overall scope of evolutionary opportunities as the hypothesized evolution of humans from our last common ancestor with chimps. The three high-confidence examples of evolution had larger numbers of organisms and stronger and more consistent selective pressures but had fewer total generations, thus leading to an *approximately* similar net opportunity for evolution.

HOW DIFFERENT ARE HUMANS AND CHIMPS?

From the previous section, the scope of the high-confidence evolution studies of malaria and *E. coli* (i.e., the degree of evolutionary opportunity) was approximately the same as that of the hypothesized evolution of humans from our last common ancestor with chimps. From chapter 7, we know the extent of evolutionary DNA changes that occurred in these three high-confidence studies.[69] Thus, if humans evolved, we should expect that the extent of evolutionary DNA change between humans and our last common ancestor with chimps

---

[69] Some readers may be thinking that my summary of the extent of evolutionary changes in the high-confidence studies has ignored the many other changes that occurred in the studies. For example, Lenski's experiment showed many other evolutionary changes that occurred in the same period of time as the ability to metabolize citrate in the presence of oxygen. In each of the high-confidence evolution studies, I focused on the direction of evolutionary change that produced the most significant difference, or the most complex evolution. In each study, other changes in the DNA certainly occurred over time. However, our interest lies in the degree of complexity of changes that evolution can produce, not in the quantity of simple changes.

would approximate the extent of evolutionary DNA changes in the high-confidence evolution studies.

We therefore need to determine the extent of the DNA changes between humans and our last common ancestor with chimps. If this is similar to that found in the high-confidence evolution studies, then the high-confidence data supports human evolution. Unfortunately, we don't have DNA from the last common ancestor, so we have to approximate the last common ancestor by way of the modern chimpanzee.

The Human Genome Project was a thirteen-year, $3 billion effort to sequence the entire human genome. The magnitude of this collaborative effort and its impact on humanity are difficult to overstate. A draft of the human genome was published in 2001, and the final result, signaling the closing of the project, was released in 2003. The final result showed that only 1.2 percent of the human genome coded for proteins. Most of the remaining DNA, as noted earlier, was considered to be junk DNA: nonfunctional remnants of our evolutionary past.

A much smaller effort was made to document the DNA of chimpanzees through the Chimpanzee Sequencing and Analysis Consortium. This group published a *draft* chimp genome in 2005.[70] Unfortunately, we do not yet have a final chimp genome. Also unfortunately, the draft chimp genome was assembled by using the physical framework of the known human genome as a "scaffold" for combining small pieces of sequenced chimp DNA. This is a type of assumption built into the draft chimp genome: one that confers a bias toward the similarity of human and chimp DNA. Comparing only the selected regions of human and chimp DNA that were sufficiently similar to allow alignment (i.e., the so-called orthologous regions), they con-

---

[70] The Chimpanzee Sequencing and Analysis Consortium. Initial sequence of the chimpanzee genome and comparison with the human genome. Nature. 2005:437;69–87.

cluded that humans and chimpanzees differ by only 1.2 percent, thus implying 98.8 percent equivalence.

This finding encouraged the now commonly held belief that chimp and human DNA are nearly identical. The comparison included only the orthologous regions, because 1) these regions are the easiest to sequence and compare, and 2) at the time of this comparison, the majority of DNA was considered to be junk DNA, and comparing regions of junk DNA seemed irrelevant at the time. As discussed in chapter 6, the more recent ENCODE study demonstrated that more than 80 percent of the human genome has known functions (compared to the original estimate of 1.2 percent). Thus we have three reasons why the level of similarity between human and chimp DNA is in need of serious revision: 1) a dramatic increase in appreciation for DNA function, 2) the use of human scaffolding to sequence chimp DNA, and 3) the comparison of human and chimp DNA only in orthologous regions.

Even if the revised percentage similarity between human and chimp DNA remains high, the immense size of the genome implies an impressive level of divergence between humans and chimps. To put this into perspective, let's consider the estimate of similarity between chimp and human DNA that the aforementioned Francis Collins (head of the Human Genome Project) offered before the ENCODE results were released. He said that "humans and chimps are 96 percent identical at the DNA level."[71] The human genome has about 3 billion base pairs of DNA. A difference of 4 percent therefore implies that humans and chimps differ by about 120 million base pairs. This is approximately the size of the entire fruit fly genome, which suggests that the extent of human evolution in the last 6 million years may approximate the extent of the fruit fly genome. The fruit fly genome is

---

[71] Collins, FS. The language of God: A scientist presents evidence for belief. New York: Free Press; 2006, p. 137.

substantially larger than the genetic changes that we observed in the high-confidence evolution studies (i.e., a handful of point mutations).

Another view of the differences between human and chimp DNA involves evidence of genes in humans that do not exist in other primates. As mentioned above, the sequencing of DNA and comparative genomics initially focused only on the segments of DNA that code for proteins and are similar across species (i.e., those that are orthologous). This is a type of homology that is commonly presented as evidence in support of evolution (see chapter 6). More recently, genes that are unique to a given species are receiving more attention. As it turns out, 10–20 percent of the genes in every eukaryotic genome have no sequence similarity to genes from other species.[72] These genes are commonly known as "orphans" or "taxonomically restricted genes." When scientists first described these genes, only a small number of species had sequenced genomes. The expectation was that similar genes would be discovered over time as the DNA of more species was sequenced. Despite exponential growth in the number of species that have been sequenced, however, the proportion of orphan genes in each species remains at 10–20 percent. This high prevalence of orphan genes came as quite a surprise to those who profess the conservation of genes over taxa as evidence for evolution. As summarized by the science writer Helen Pilcher in the British magazine *New Scientist*:

> When biologists began sequencing genomes, they discovered that up to a third of genes in each species seemed to have no parents or family of any kind. Nevertheless, some of these "orphan genes" are high achievers, and a few even seem to have played a part in the evolution of the human brain.

---

[72] Khalturin, K, et al. More than just orphans: Are taxonomically-restricted genes important in evolution? Trends in Genetics. 2009;25:404–13.

But where do they come from? With no obvious ancestry, it was as if these genes had appeared from nowhere, but that couldn't be true. Everyone assumed that as we learned more, we would discover what had happened to their families. But we haven't—quite the opposite, in fact.[73]

If 10–20 percent of the genes in any given species are unique to that species, that implies that the remaining 80–90 percent of the genes are shared or conserved across species. Imagine a database that contains a list of all the genes of all living organisms. The first organism would have each of its genes entered into the database. The second organism would have its orphan genes entered, but the shared genes already appear in the database. The third organism would add more orphan genes but no additional shared genes, and so on. Because the unique orphan genes are each listed once in the database, and the shared genes are also listed only once, this suggests that upon completion of the database we would find the subset of orphan genes to be larger than the subset of shared genes. A larger overall number of unique genes than shared genes should lead us to question the paradigm of evolution from a common ancestor. This is the kind of evidence that is never mentioned when homology is used as an argument for evolution.

The study of human-specific orphan genes is still in its infancy, in part because the final chimpanzee genome is not yet available, and in part because studies of orphan genes have almost exclusively focused on protein-coding genes. Nevertheless, many reports of human-specific genes have been published. Recent sequencing of the chimpanzee Y chromosome was completed with use of a chimp-specific scaffold (i.e., not using the human scaffold as was the case in the draft chimp

---

[73] Pilcher, H. All alone. New Scientist. January 19, 2013;38–41.

genome).[74] This resulted in a conclusion that is perhaps best stated by the title of the paper by the geneticist Jennifer Hughes and colleagues: "Chimpanzee and Human Y Chromosomes Are Remarkably Divergent in Structure and Gene Content." Looking at the portion of the Y chromosome that is male-specific, the study authors found that humans had seventy-eight genes, whereas chimps had only thirty-seven. Numerous other publications have agreed that the number of human-specific orphan genes is large and that the number will continue to grow as we learn more. The following list of summary statements from these publications supports a surprising level of difference between humans and chimps.

Our results imply that humans and chimpanzees differ by at least 6 percent (1,418 of 22,000 genes) in their complement of genes.[75]

For about 23 percent of our genome, we share no immediate genetic ancestry with our closest living relative, the chimpanzee. This encompasses genes and exons to the same extent as intergenic regions.[76]

Here we identify 60 new protein-coding genes that originated de novo on the human lineage since divergence from the chimpanzee.[77]

---

[74] Hughes, JF. Chimpanzee and human Y chromosomes are remarkably divergent in structure and gene content. Nature. 2010;463:536–9.

[75] Demuth, JP, et al. The evolution of mammalian gene families. PLoS One. 2006;1:e85.

[76] Ebersberger, I, et al. Mapping *Human Genetic Ancestry*. Mol Biol Evol. 2007;4:2266–76.

[77] Wu, D-D, et al. De novo origin of human protein-coding genes. PLoS Genet. 2011;7:e1002379.

It seems prudent to conclude that substantial changes occurred in the human gene reservoir with about 300 human-specific genes and 1000 primate-specific genes added.[78]

More than 6 percent of genes found in humans simply aren't found in any form in chimpanzees. There are over fourteen hundred novel genes expressed in humans but not in chimps.[79]

The last statement is perhaps the most unexpected, because it is taken from Jerry Coyne's book that is intended to provide airtight proof for macroevolution. In the midst of a very large quantity of low-confidence evidence for macroevolution, he pauses to express his surprise at the large degree of genetic difference between humans and chimpanzees. He offers no explanation for how the 1,400 novel human genes came to be and then quickly returns to his low-confidence evidence for human evolution.

The average size of a human gene is about 54,000 base pairs of DNA. Thus, 1,400 new genes would require approximately 75.6 million base pairs of new information added to human DNA. This estimate is not far from the 120 million base pairs from our first estimate (~4 percent of the human genome and the size of the fruit fly genome).

Let us now return to how this section started. We said: "If humans evolved, we should expect that the extent of evolutionary DNA change between humans and our last common ancestor with chimps would approximate the extent of evolutionary DNA changes in the

---

[78]   Zhang, YE, et al. New genes contribute to genetic and phenotypic novelties in human evolution. Current Opinion in Genetics & Development. 2014;29:90–6.

[79]   Coyne, JA. Why evolution is true. London: Penguin Books; 2009, p. 211, referring to the study of Demuth et al.

high-confidence evolution studies." It is time to compare results. **The development of resistance to chloroquine required between four and eight base pair changes, whereas human evolution from our last common ancestor would require approximately 75 million base pairs of new DNA.** You should now be able to see that the rough estimates of the scope of evolutionary opportunity (like the total number of humans/hominids ever to exist) are not of critical importance, since the gap is a factor of 10 million, not a factor of 10, 100, or 1000. To develop resistance to chloroquine, malaria required 100 million times the number of humans/hominids that ever existed. Yet, the putative evolution of humans from our last common ancestor involves 10 million times the amount of DNA modification that malaria required to develop resistance to chloroquine.

The immense difference in the extent of evolutionary changes is obvious, but, as we learned from chapter 7, these simple numbers fail to provide an appreciation for the challenges in evolving new genes. Perhaps you will recall the following (rather lengthy) sentence from the "Tryptophan Production" section of chapter 7:

> Given that the alpha subunit of the tryptophan synthase gene is 268 amino acids (804 base pairs), and given that the beta subunit is also necessary for producing tryptophan (the beta subunit is 397 amino acids or 1,191 base pairs), it is sobering to try to imagine a pathway where each single mutation along the path to the functional enzyme could confer a fitness benefit (even if portions are "borrowed" from existing genes) and where no energy-saving sidetrack leads the evolutionary process astray.

Our discussion about human evolution adds a whole new dimension of sobriety to this statement. Despite the fact that high-confidence studies of evolution have failed to produce any new genes (and have even failed to make minor repairs to an existing gene), and even though they had similar evolutionary opportunity as the putative evolution of humans from our last common ancestor with chimps, we are to believe that at least 1,400 new genes were produced in the evolution of humans, where each single mutation along the path to each of the 1,400 new genes conferred a fitness benefit and where no energy-saving sidetracks led the evolutionary process astray.

THE RATIONAL CONCLUSION

We have reviewed the commonly cited evidence for evolution and have applied well-recognized criteria to determine the level of confidence afforded by each line of evidence. Although the criteria for levels of confidence are not black and white, but rather include many shades of gray, we found that the evidence for evolution is clearly dichotomous. The set of high-confidence evidence for evolution includes prospective studies on organisms that reproduce rapidly and in great quantities. These studies clearly demonstrate microevolution, with up to eight mutations working together to confer benefits under specific selective environments. We then applied the high-confidence studies to test the concept of macroevolution, focusing on the hypothetical evolution of humans from our last common ancestor with chimpanzees, starting 6 million years ago. We showed that the scope of evolutionary opportunity was similar to that found in the high-confidence studies. The extent of evolutionary changes between humans and our last common ancestor with chimps should therefore approximate the extent of evolutionary changes seen in the high-confidence studies (i.e., up to eight mutations working together to confer a benefit in any particular direction). Finally, we reviewed the known genomic differences between humans and chimps (with modern chimps used to approximate our last common ancestor with chimps) and uncovered more than 1,400 novel genes—approximately 75 million base pairs of new DNA—in humans. Although more research is clearly needed, **the conclusion of our work (presented with sober judgment) is that the evidence for evolution that was obtained by high-confidence science casts significant doubt on claims of human evolution from our last common ancestor with chimpanzees.**

The above conclusion is offered with recognition of several limitations. First, the chimp genome was used to represent the genome of our putative last common ancestor. This assumption is commonly made out of necessity, especially if there was no common ancestor. We don't know the genome of the common ancestor, but if more than 1,400 genes separate humans and chimps, and if we accept that humans represent an advancement from our last common ancestor, we can be confident that at least hundreds of genes separate humans and the common ancestor.

Second, the chimp genome is not yet complete, and many more discoveries are likely to be made. But the best-sequenced portion of the chimp genome (the Y chromosome) showed substantial differences between chimp and humans, which suggests that completion of the chimp genome will only reveal greater differences between chimp and human.

Third, the comparison of the scope of evolutionary opportunities between the high-confidence studies and human evolution required a rough estimate of the total number of humans and hominids ever to have existed. This comparison is intended to be an approximation; it is not intended to be presented with a high degree of accuracy. The estimate of 1 trillion could easily be off by a factor of 10 or 100 but is certain to remain well below the estimated number of malaria organisms needed to develop resistance to chloroquine (one hundred million times one trillion).

Finally, this argument involves applying lessons learned from microbe evolution and applying them to hominids. One could claim that the differences between these organisms could change the process of evolution and invalidate the result. The microbes included both prokaryotes and eukaryotes, however, which are vastly different but have provided similarly constrained views of evolution.

9

It Is Called "Faith"

In the chapter on low-confidence evidence for evolution, I held back the aspects of the grand definition of evolution that have the lowest possible scientific confidence. As we will see, science actually has no applicability to these particular claims of the grand definition of evolution, and the implications are profound.

According to a Gallup poll in 2014, 19 percent of Americans believe that "Human beings have developed over millions of years from less advanced forms of life, but God had no part in this process." If "God had no part in this process," that implies that God had no part in producing the very first living organism. In other words, life spontaneously arose from nonlife, a hypothetical process called "abiogenesis." Darwin himself suggested this:

> It is often said that all the conditions for the first production of a living organism are now present, which could ever have been present—But if (and Oh! what a big if!) we could conceive in some warm little pond with all sorts of ammonia and phosphoric salts,—light, heat, electricity etc., present, that a protein compound was chemically formed, ready to undergo still more complex changes, at the present day such matter

would be instantly devoured, or absorbed, which would not have been the case before living creatures were formed.[80]

Darwin suggested the formation of a protein that was "ready to undergo still more complex changes." Although we know that he was talking about the origin of life, all will agree that a protein, even a complex one, is not a living thing. Since the time of Darwin, the possibility of spontaneous generation of life has been hotly debated. Eugene Koonin, a firm believer in grand evolution and an expert on the genetics and molecular biology of microorganisms, has this to say about abiogenesis:

The origin of life is the most difficult problem that faces evolutionary biology and, arguably, biology in general. Indeed, the problem is so hard and the current state of the art seems so frustrating that some researchers prefer to dismiss the entire issue as being outside the scientific domain altogether, on the grounds that unique events are not conducive to scientific study.[81]

Indeed, it is hard to imagine that science could adequately address a hypothesized unique event that happened 4 billion years ago. Let's ponder the application of science to answer the following question: Did the first life spontaneously arise from nonlife? Of all the questions that may provoke the application of the tool of science, this question ranks among the least appropriate. In other words, the tool of science has no business trying to answer this question. The table below explains why this is such an extreme of low-confidence science.

---

[80] From Darwin's letter to the botanist and explorer Joseph D. Hooker, February 1, 1871.
[81] Koonin, EV. The logic of chance: The nature and origin of biological evolution. Upper Saddle River, NJ: Pearson Education, Inc; 2012, p. 351.

Criteria of Low-Confidence Science	How they apply to the question: Did the first life spontaneously arise from non-life?
1. Can't be repeated	Even if a laboratory setting could demonstrate life arising from non-life (a result that has never been observed), one could not be confident that the same conditions existed for the original hypothetical abiogenesis event, about 4 billion years ago.
2. Indirectly measured, extrapolated, or inaccurate results	A hypothetical abiogenesis event has never been observed. The most extreme form of indirect measurement is something that cannot be measured!
3. Retrospective, observational study	Something that supposedly occurred 4 billion years ago certainly counts as retrospective. This cannot even be considered an observational study, because no credible evidence remaining from this hypothetical event is observable.
4. Clear opportunities for bias	With nothing to measure and no data, bias is all that remains.
5. Many assumptions required	With nothing to measure and no data, assumptions are essential. The most common assumptions are naturalism (i.e., absence of any supernatural intervention), and directly assuming that abiogenesis occurred: because there is life and supernatural intervention is excluded, it must have started spontaneously.
6. Overstated confidence or scope of results	Any claim that science supports or even suggests abiogenesis is fantasy.

Let's ask a more fundamental question: Is it possible to prove that abiogenesis *did not* happen? Anyone can come up with a theory. Incorrect theories need to be excluded by following a scientific process. If it is not possible to show that a theory is false (i.e., if you cannot conceive of a method that could demonstrate that the theory is false), it cannot be scientific, because it can never be excluded. Falsifiability (the possibility of proving that something is false) is a well-known requirement for a theory to be scientific. From a current biology textbook: "It is essential in science that incorrect ideas are discarded, and that can occur only if it is possible to prove those ideas false."[82] Jerry Coyne also agrees: "If you can't think of an observation that could disprove a theory, that theory simply isn't scientific."[83]

Because we are on the topic of abiogenesis, you might think that these quotes were made in reference to abiogenesis. In fact, both of these quotations were speaking of supernatural explanations for life (i.e., creation). A belief in supernatural creation requires faith, so I have no argument with applying these statements to creation. But I do have a problem if this philosophy of falsifiability is applied only to supernatural explanations for life and abiogenesis is somehow exempt. One would hope that their stated philosophy of science wouldn't change if the subject matter changed from creation to abiogenesis.

I challenge anyone to think of a test or an observation that could prove that abiogenesis *did not* happen. Simply put, **the theory that the first life spontaneously came from nonlife is not falsifiable and therefore is not scientific.** Unfortunately, neither the biology textbook nor Jerry Coyne recognize this. They are happy to use their

---

[82] Belk, C, and Borden Maier, V. Biology: Science for life (with Physiology). Third edition. San Francisco: Benjamin Cummings; 2010, p. 3.

[83] Coyne, JA. Why evolution is true. London: Penguin Books; 2009, p. 138.

falsifiability argument against their opponents as long as it doesn't apply to their own beliefs.

The fact that abiogenesis is not scientific has profound implications. It implies that the origin of first life is not a topic that science can address. If science cannot be used to support a position, and yet people hold to a strong belief in that particular position (either a "pro" or "con" position), there is only one word to describe their belief: *faith*. A common definition of faith is "a belief that is not based on proof." Another well-known definition of faith is "confidence in what we hope for and assurance about what we do not see." No one has ever seen life come from nonlife. Some may say: "Yes, but no one has ever seen gravity, yet we know that it exists." Although it is true that gravity cannot be seen (by the human eye), it can clearly be "seen" (i.e., measured) during high-confidence experimentation such as the dropping of a bowling ball. We're talking about a very generalized definition of "seen," but, even using the most generalized definition, no one has ever seen life come from nonlife.

Those who have faith in abiogenesis usually bring up the well-known experiments of Stanley Miller and Harold Urey (and others like them). Miller and Urey were able to produce individual amino acids in a lab environment that was intended to simulate the conditions of a young Earth. Unfortunately, their experiment, and all similar experiments, resulted in a mixture of left-handed and right-handed amino acids. (The terms "left-handed" and "right-handed" refer to "chirality": two forms of the same molecule that are mirror images and are not superimposable.) As it turns out, all life on Earth exclusively uses left-handed amino acids. The introduction of right-handed amino acids would interfere with protein function. As a result, those who have faith in abiogenesis must imagine that some natural filter excluded the right-handed amino acids. Living organisms have such filters that select only the left-handed amino acids, but those structures are built, in part, by proteins that are made of exclusively left-handed amino acids. Even if you could filter out the right-handed amino acids, nobody believes that amino acids are alive. If amino acids form into a chain called a protein, the function and complexity of a protein remains a long, long way from that of a living organism. A living organism requires 1) a source of information (like DNA or RNA) to code for the production of macromolecules, 2) a means of metabolism (i.e., a way to derive chemical energy to build structures or to combat degradation), and 3) the ability to reproduce itself. Although RNA molecules can be shown to perform portions of these three functions (part of the "RNA World"— a hypothesis that RNA molecules were the precursors to all life on Earth), no single molecule can perform all three, and RNA is quite fragile. Multiple molecules that work together to perform these functions would require a containment system (i.e., a cell) to keep the process of diffusion from distancing the essential molecules. In short, the RNA World hypothesis is wishful thinking.

In existing forms of life, the source of information (DNA) is decoded by a set of molecules (proteins) to instruct another set of proteins to build the necessary molecules for metabolism, cell structure, and reproduction. We know, however, that the proteins are needed to reproduce the DNA, and the DNA is needed to produce the proteins. This interdependency begs the question of which came first.

We may then ask: What is the minimum complexity required for life? Looking at the existing forms of life, scientists point to the bacterium *Mycoplasma genitalium* as the simplest example of an independent free-living organism. (Simpler organisms do exist, but they are dependent on other organisms to survive—in other words, they are parasitic.) *Mycoplasma genitalium* contains 517 protein-encoding genes. Scientists believe that between 265 and 310 of these 517 genes are absolutely required for the organism to survive and reproduce.[84] A self-contained organism with 265 genes is vastly more complex than the mixture of left-handed and right-handed amino acids produced by Miller and Urey or the small chains of amino acids or RNA produced by similar experiments. Claiming that the spontaneous production of these molecules supports the idea of the spontaneous generation of life from nonlife is like claiming that the production of a single transistor supports the idea of the spontaneous generation of an iPhone: except that an iPhone is not capable of reproducing itself, whereas life must be able to reproduce itself. The experiments by Miller and Urey, and others like them, somehow give people a degree of confidence in what they hope for: abiogenesis. The only appropriate word for this is "faith."

The aforementioned biology textbook, and five other biology textbooks I surveyed, discussed the Miller/Urey experiment and others like it, presenting only one side of the discussion: the sparse data that could encourage a belief in abiogenesis. The textbook authors made no mention of the limitations we've described above (left-handed amino acids, the interdependency of proteins and DNA, and the immense complexity of the simplest known life-form). One must wonder why they would only present encouraging results and not discuss the limitations. Belk and Borden Maier's textbook concluded that "Although life as we would identify it has not been created in the lab from scratch, these results support the hypothesis that life could have been formed spontaneously on Earth."[85] The introductory chapter of this same biology textbook espouses falsifiability as a require-

---

[84] Hutchison, CA III, et al. Global transposon mutagenesis and a minimal mycoplasma genome. Science. 1999;286:2165–9.

[85] Belk, C, and Borden Maier, V. Biology: Science for life (with Physiology). Third edition. San Francisco: Benjamin Cummings; 2010, p. 248.

ment for a theory to be scientific. If they would have adhered to their own stated doctrine of falsifiability, this concluding statement would be exposed as a statement of faith. Their concluding statement should actually be: "Despite the inability to demonstrate abiogenesis experimentally and despite the fact that abiogenesis is not scientific, we have presented our best argument to encourage you to maintain faith that life arose spontaneously."

Whenever I explain to someone in this camp that abiogenesis is based on faith, they simply cannot imagine that their belief is not based on science.[86] I suppose that if science is your god, then you will be unable to see its limitations. Bill Nye wrote a chapter about abiogenesis called "The Sparks That Started It All," where he states: "The origin of life just requires some raw material that could allow the spark of life to emerge."[87] This is a statement of faith, not science. Claiming that this has anything to do with science does a tremendous disservice to science. To paraphrase Nye's own words against him,[88] if you want to believe that this is science, that's fine. But please do not teach this to your children—we need them to be good scientists and engineers. Whether you believe that life spontaneously came from nonlife or that life came from the will of a divine being, your belief about the origin of life requires faith, because science does not apply.

The rational conclusion that belief in abiogenesis is a demonstration of faith is sure to make many people uncomfortable, especially

---

[86] At the time of this writing, the top of the Wikipedia page for abiogenesis says, "For non-scientific views on the origins of life, see *Creation myth*." This implies that abiogenesis, in contrast to creation, is scientific. I beg to differ.

[87] Nye, B. Undeniable: Evolution and the science of creation. New York: Saint Martin's Press; 2014, p. 285.

[88] https://www.youtube.com/watch?v=gHbYJfwFgOU.

the 19 percent of Americans who believe that "human beings have developed over millions of years from less advanced forms of life, but God had no part in this process." The word "faith" is repulsive to many in this camp:

> *Faith is the great cop-out, the great excuse to evade the need to think and evaluate evidence. Faith is belief in spite of, even because of, the lack of evidence.*
>
> —RICHARD DAWKINS

> *It is time that we admitted that faith is nothing more than the license religious people give one another to keep believing when reasons fail.*
>
> —SAM HARRIS

> *It is called faith because it is not knowledge.*
>
> —CHRISTOPHER HITCHENS

We can now see the hypocrisy here: people who revile faith are actually people of faith. Some in this group have recognized this and have attempted to separate abiogenesis from grand evolution. They are attempting to build a wall between the aspects of their worldview that are faith-based and those that are science-based. But it is not so simple. Abiogenesis isn't the only part of grand evolution that requires faith. We previously mentioned that all of life on Earth is organized into three domains (the highest taxonomic rank): Archaea and Bacteria (together called prokaryotes) and Eukarya. Eukaryotes have much larger cells and very much more complex cell structures and DNA than prokaryotes. Eukaryotic cells have a nucleus and organelles, whereas prokaryotes are relatively disorganized. In prokaryotes, the cell contents freely mix in a solution by simple diffusion. In contrast, eukaryotes are compartmentalized. Organelles bound by membranes keep tight control on what gets in and what gets out. Macromolecules cannot get from "point A" to "point B" by simple diffusion—they are transported by complex trafficking systems. In prokaryotes, DNA is freely accessible. In eukaryotes, DNA is found inside the nucleus, and highly complex nuclear pores control what gets in and out of the nucleus. These are but a small sample of the vast differences between prokaryotes and eukaryotes. You could think of prokaryotic cells as bicycles and eukaryotic cells as Tesla Model S electric cars.

The grand definition of evolution implies that eukaryotes evolved from prokaryotes. The origin of eukaryotes (a process called "eukaryogenesis") is a profound mystery. In his recent book *The Logic of Chance*, the aforementioned Eugene Koonin speaks of the dramatic increase in complexity of eukaryotes over the other domains:

> No direct counterparts to the signature eukaryotic organelles, genomic features, and functional systems exist in archaea or bacteria. Hence, the very nature of the evolutionary relationships between prokaryotes and eukaryotes becomes a cause of bewilderment. [...]

> Of the three domains of life, eukaryotes possess by far the most complex, strikingly elaborate cellular organization that for some might even summon the specter of "irreducible complexity" because for most of the signature functional

systems of the eukaryotic cells, we can detect no evolutionary intermediates.[89]

We have determined that abiogenesis is not falsifiable. Similarly, I challenge anyone to think of a test or an observation that could prove that eukaryotes *did not* evolve from prokaryotes, or that eukaryotes *did not* arise spontaneously from nonliving matter. Thus, one's belief in how eukaryotes came to be is also a matter of faith, not science. Here you can see the development of a slippery slope: Where does faith end and science begin?

Those who have faith in eukaryogenesis will be quick to provide evidence for what is known as endosymbiosis. This is a theory that small prokaryotes began to live within larger cells and evolved to perform specific functions for the larger cell, thus developing a symbiotic relationship. The claim is that mitochondria (the energy-producing organelle found in eukaryotic cells) formerly were separate prokaryotic cells. The biology textbooks again focus on this hypothesis to support a belief in eukaryogenesis but fail to mention the chasm that remains between prokaryotes and eukaryotes. Endosymbiosis somehow gives people confidence in what they hope for: eukaryogenesis. The only appropriate word for this, as noted earlier, is faith.

To summarize this chapter, one's worldview (grand evolution or creation) can only be based on science or faith. Those who believe in grand evolution are eager to show that science is their foundation and that science has provided proof. As we have seen, with sober judgment, faith lies at the foundation of grand evolution. Similarly, faith is also the foundation of a belief in creation.

---

[89]   Koonin, EV. The logic of chance: The nature and origin of biological evolution. Upper Saddle River, NJ: Pearson Education, Inc; 2012. First quote: p. 174. Second quote: p. 221.

Francis Collins's book *The Language of God: A Scientist Presents Evidence for Belief* includes a chapter on creationism with the subtitle "When Faith Trumps Science." By this, he is implying that grand evolution creates a conflict between faith and science for anyone who believes in creation and that creationists have elected to subordinate science to their faith. As we have just discussed, however, the foundations of grand evolution (abiogenesis and eukaryogenesis) cannot be addressed by science; they are left to the realm of faith. Rather than creating a conflict between science and faith, the question of the origin of life thus creates a conflict between two faiths: faith in creation versus faith in abiogenesis and eukaryogenesis. In both worldviews, faith trumps science, because science has nothing to say; it is not the right tool to address the origin of life.

In addition, we reviewed the available high-confidence evidence for evolution in chapter 7 and applied it to the hypothetical evolution of humans from our last common ancestor with chimpanzees in chapter 8. We found that science does not support this claim of evolution. For creationists, their belief that humans did not evolve from other primates is therefore supported both by science and by their faith. Again, there is no conflict between science and faith. Only the low-confidence evidence, which is obtained by stretching the tool of science beyond its limits, seems to create a conflict between science and faith. **Rather than having science trump faith or faith trump science, the appropriate application of science, with a healthy appreciation of its inherent limitations, resolves the perceived conflicts. Scientists should therefore apply science appropriately to gain knowledge for the benefit of humanity rather than stretch it beyond its limits to create artificial conflict.**

The last paragraph of chapter 1 said, "This book offers a new approach to reconciliation." The new approach is now clear: it comes

from taking the higher scientific ground through the appropriate application of science and an appreciation of its limitations. That is '*The
Scientific Approach to Evolution*'.

One's belief about the origin of first life is a matter of faith; it is
not a question that can be addressed by science. Therefore, the origin of life should not cause any conflict between science and faith.
High-confidence science has clearly demonstrated microevolution.
Microevolution should therefore be accepted both by those who believe in the grand definition of evolution and by those who believe
in creation. For this reason, microevolution should not cause any
conflict between science and faith or between creationists and evolutionists. Agreement from both sides that the origin of life is a matter
of faith and that microevolution is undeniable would go a long way
toward reconciliation.

The remaining focus of conflict lies in macroevolution—an area
that is difficult to study with the tool of science. The study of macroevolution using the tool of science runs the risk of stretching science
beyond its limitations and leading to the adoption of fallacies. The
only support for macroevolution comes from one interpretation of the
very-low-confidence evidence, summarized in chapter 6. Creationists
apply a different interpretation of the same very-low-confidence evidence to refute macroevolution (for example, the discontinuities of
the Cambrian explosion, the relative paucity of transitional forms in
the fossil record, and the large gaps between taxonomic classifications
of existing life forms). This conflict occurs because the very-low-confidence evidence does not provide clarity; it only provides opportunity for the influence of bias. There is no high-confidence evidence
of macroevolution; there is only high-confidence evidence of microevolution. The high-confidence evidence of microevolution provides
insights into the limited capacity of evolution. The application of

the high-confidence evidence of microevolution to the study of macroevolution (chapter 8) casts significant doubt on the likelihood of macroevolution. Clearly, additional work is needed here—work that respects the limitations of science and prioritizes higher-confidence approaches. If we focus on higher confidence approaches and take the higher scientific ground in continuing to study macroevolution, I believe that reconciliation here is also possible.

10

The Scientific Higher Ground

It's hard to argue with the need for good education. Education is one of the keys to a brighter future for humanity and for our planet. Education is also the motivation behind this book. My appreciation for the differences between high- and low-confidence science has largely developed through nearly twenty years of experience as a scientist—not through formal education. My years of schooling, including ten years of higher education in engineering and medicine, did not convey a clear appreciation for the criteria of high-confidence science. In other words, the curriculum did not emphasize the importance of repeatability, directly measured results, prospective interventional study, the avoidance of bias and assumptions, and the sober presentation of results in the practice of science.

As my children progressed through public school education, I hoped they would receive a better science education than I had. Their high school biology class was a tipping point for me. The evolution curriculum failed to appreciate the differences between high-confidence and low-confidence science, failed to mention any limitations or assumptions behind the evidence, and failed to acknowledge that abiogenesis and eukaryogenesis cannot be addressed by science. Macroevolution was presented as a simple and inevitable progression of microevolution and only low-confidence evidence (exclusively in support of macroevolution) was mentioned. There was no mention of the possibility that evolution is constrained, as suggested by the high-confidence evidence in chapter 7 and the discussion of chapter 8. Although the biology textbooks admitted that abiogenesis had not been proven, they provided their best arguments to support a belief in abiogenesis. In short, the curriculum for evolution was a partisan effort to indoctrinate students into accepting grand evolution rather than an open discussion of the strengths and limitations of the available evidence. There was no sober judgment of results, no statement of assumptions or limitations, no disclosure of bias, and no appreciation for the level of confidence in the evidence. In the field of medicine, a research article written in this style would have great difficulty being published in a reputable medical journal. A marketing brochure would be a more appropriate genre for the current evolution curriculum in public schools.

We must ask: Why is evolution taught in this manner? Many future doctors, engineers, and scientists are being educated in this manner each year. Shall we teach them to prefer low-confidence evidence; to ignore any limitations, bias, and assumptions; and to only consider the evidence that supports their view?

I believe that the current approach to teaching evolution in public schools has resulted from a combination of several factors. First, as we discussed in chapter 9, any explanations for life that involve a supernatural intervention cannot be falsified, and therefore cannot be studied by science. I have no disagreement here. Second, the separation of church and state mandates that religion (i.e., supernatural interventions) cannot be taught in public schools. Again, I have no disagreement here. These two statements are not the cause of the problem. However, these two statements, combined with the bias of those who control the curriculum, have led to an illogical conclusion:

any possibility of supernatural intervention must be excluded from explanations of life. This is the assumption of naturalism: everything arises only from natural properties and causes.

The assumption of naturalism, and the suggestion that it is supported by science, creates four major problems in the current evolution curriculum in public schools, as follows.

1. As discussed in chapter 9, the naturalistic explanation for the origin of life (i.e., abiogenesis) cannot be addressed by science, just as a supernatural explanation for the origin of life cannot be addressed by science. Yet biology textbooks make a clear effort to encourage belief in abiogenesis rather than to admit this limitation.

2. Supernatural intervention is not falsifiable; everyone seems to agree on that. But the assumption of naturalism can only be made if supernatural intervention is falsified, which is not possible. Thus, between 1) and 2), we have something that cannot be falsified (supernatural intervention) being treated as false and something else that cannot be falsified (abiogenesis) being treated as true in the current evolution curriculum.

3. By excluding any possibility of supernatural intervention, the biology curriculum in our public schools has taken free license to make a one-sided appeal for naturalism and grand evolution as the scientific approach to origins. By claiming that naturalism is supported by science, students are encouraged to abandon their "nonscientific" religious beliefs and to adopt atheism. An effort to abide by the separation of church and state (by avoiding any discussion of supernatural intervention) has thus been turned into an infraction of the separation of church and state (by encouraging students to abandon their current beliefs).

4. The current biology curriculum discredits itself by refusing to admit the limitations of science, by prioritizing low-confidence science over high-confidence science, and by subordinating the scientific method to the promotion of an agenda. This damages the student's appreciation of the scientific method and its limitations.

How, then, can evolution be taught properly in public schools without discussing supernatural intervention? By taking the scientific higher ground—that is, by teaching the scientific approach to evolution. Taking the higher ground sometimes involves admitting limitations rather than practicing overcompensation to conceal those limitations. For example, I've never piloted an airplane. If I were on an airplane and the pilot and copilot suddenly died, and one of the flight crew asked the passengers if anyone knew how to fly the plane, I could practice overcompensation by boldly stating (in a deep and courageous voice), "Yes, I will be happy to land this plane." Alternately, I could admit my limitation by humbly stating that I have never flown an airplane; if anyone else has experience, that person would be a better choice. Which would lead to a better outcome? The same is true for science. Open discussions of the limitations of science will lead to a greater appreciation of the tool of science and greater interest in applying high-confidence science to the benefit of society. The scientific approach to teaching evolution involves

- teaching an appreciation for the difference between high-confidence and low-confidence science and an appreciation for the tool of science: both its strengths and its limitations;
- teaching, with appropriate humility, that the tool of science is not applicable for addressing the origin of first life. As a

result, one's belief in how life began is up to each individual and always will be. The origin of life lies in the realm of faith, not science. This is no different than the subject of life after death. Science is not applicable to addressing life after death, and the subject of life after death is not a part of public school education. Why, then, is the origin of life a part of public school education?

- demonstrating the appropriate prioritization of high-confidence evidence over low-confidence evidence;

- presenting the high-confidence evidence of microevolution and showing that it has practical implications for our society, such as the development of resistance to antibiotics;

- explaining the scope of genetic changes that occur in high-confidence experiments on evolution, even when very large numbers of organisms and generations are included. This high-confidence evidence suggests that the process of evolution may be constrained; and

- teaching an appreciation for the difficulty in applying science to macroevolution by showing that the commonly cited evidence is low-confidence evidence. As a result, there are conflicting interpretations of the low-confidence evidence; for example, aspects of the fossil record suggest transitional forms, while other aspects of the fossil record suggest a combination of stasis and sudden discontinuities. Here, the tool of science is being stretched beyond its limitations, thus opening the door for conflicting interpretations and bias.

Isn't it more open and honest to teach our students in this manner rather than degrading their understanding and appreciation for science or to encourage them to abandon their religious beliefs via the current curriculum? The approximately 5 million teenagers who take biology each year in public schools in the United States deserve better. We live in a world where cans of nuts must openly disclose the low confidence about their claims (Figure 2 of chapter 2), but our biology curriculum is filled with low-confidence science and yet boldly claims high confidence in grand evolution. It is time to take the scientific higher ground in biology education.

It is also time to take the scientific higher ground in the practice of science. Approximately 70 percent of medical journal articles openly disclose their study limitations. Why is this not 100 percent, and why is this practice so rare in other fields such as paleontology?[90] Whenever possible, the practice of science should adopt high-confidence approaches, and more high-confidence studies of evolution should be conducted. When it is not possible to practice high-confidence science, authors should provide justification for their approach or list their approach as a limitation, being careful to provide sober judgment of the scope of their results given these limitations.

Finally, it is time to take the scientific higher ground in our museums and media. The Google Doodle mentioned in chapter 6, portraying Lucy as a direct ancestor to humans, is a very common icon of evolution. We've seen that only very-low-confidence evidence would suggest that Lucy is an ancestor to humans (chapter 6). We have also seen that high-confidence evidence for evolution casts doubt on human evolution from primates (chapter 8). But because of its ubiquity, this icon of low-confidence science has become engrained in our minds.

---

[90] Over concerns that I might be labeled a hypocrite, I reviewed the seventeen peer-reviewed papers that I have authored. Nine of the seventeen (53 percent) openly disclose study limitations, but all six written in the last decade include open disclosure of the study limitations. Also note that the end of chapter 8 lists the limitations of applying the high-confidence evidence of microevolution to the putative evolution of humans from primates. Thus, I have learned my lesson (although I did not learn it during my formal education).

In the Vatican, the Sistine Chapel displays some of Michelangelo's greatest work. Perhaps the most famous image from the chapel is God stretching out his hand to give life to Adam. As people view this image and the specific portrayal of its figures, they unquestionably view it as a combination of imagination and art to represent a conceptual event. Believing that the event actually occurred as portrayed is clearly a matter of faith; it is not a situation where science can be applied. This art would be quite out of place in a museum of science.

FIGURE 11: *The Creation of Adam* in the Sistine Chapel. (Michelangelo)

Like Michelangelo, John Gurche is a highly acclaimed artist whose works include a combination of imagination and art. Unlike Michelangelo, Gurche's art is influenced by a collection of bones and appears in science museums, including the Hall of Human Origins in the Smithsonian Museum of Natural History. His works are viewed by millions of people each year who are interested in learning about science. One famous diorama (Figure 12) depicts Lucy. I have no reservations with an artist attempting to reconstruct a fossil organism, as long as the assumptions and limitations are disclosed. However, placing this art in the Hall of Human Origins makes a clear statement that

Lucy is an evolutionary ancestor to human beings. There is no stated recognition of faith or imagination, no disclaimers, and no disclosure of limitations presented in the display.

Thus, we have a stark contrast between two works of art: both address human origin and both were created by talented artists, but one work is appropriately recognized as art whereas the other is inappropriately positioned as a scientific explanation for the origin of humans in a museum of science. Children who visit this display believe that they are experiencing a high-confidence scientific result. I'm hoping that at least some readers will be disturbed by this dichotomy. Isn't it time that we take the higher ground in science?

FIGURE 12: Lucy in the Hall of Human Origins at the Smithsonian Museum of Natural History. (John Gurche)

Appendix A

Hierarchy of Scientific Evidence

Science has delivered enormous value to society by improving the practice of medicine. One could argue that medicine is the single-most beneficial application of science. People literally trust their lives each day to the science of medicine. If you were to become gravely ill, I'm sure you would prefer to have treatments that are supported by the highest-confidence evidence. As a result, the field of medicine has developed a strong appreciation for the six criteria of high-confidence science.

The field of medicine learned many of its lessons the hard way, with countless thousands of patients suffering from quack doctors and unproven therapies. Some of the best-known examples include snake oil salesmen,[91] leeches for bloodletting, lobotomy for mental illness (often with the use of an ice pick and no anesthesia), electromagnetic therapy or shock therapy for just about any illness, the curative power of radio-

---

[91] The original snake oil came from Chinese water snakes, imported by thousands of Chinese migrant laborers in the middle of the nineteenth century. It had legitimate and well-recognized benefits. Because it had a great reputation, imitators such as Clark Stanley took advantage of the business opportunity to sell fraudulent products. He was fined twenty dollars for violation of the Food and Drugs Act.

active water, and thousands of tonics, tinctures, balms, salves, syrups, and hair-restoration formulas. Society eventually realized that people needed to be protected from these practices. Government agencies and laws were established to regulate the sale of medical products, such that new products had to be supported by high-confidence science. In 1906, the Pure Food and Drugs Act was passed in the United States, starting the activities of what is now called the Food and Drug Administration (FDA). Today, medical products sold in the United States must first receive approval from the FDA. The FDA website says:

> FDA is responsible for protecting the public health by assuring the safety, efficacy and security of human and veterinary drugs, biological products, medical devices, our nation's food supply, cosmetics, and products that emit radiation.

> FDA is also responsible for advancing the public health by helping to speed innovations that make medicines more effective, safer, and more affordable and by helping the public get the accurate, science-based information they need to use medicines and foods to maintain and improve their health.[92]

The FDA faces a tough balance: advancing public health by speeding innovations (i.e., fast approval of new technologies) while ensuring the safety and efficacy of drugs and medical devices (i.e., requiring high-confidence scientific data on safety and beneficial effect). In my career in medical devices, I often had to negotiate with the FDA on the quantity and quality of evidence that was required to introduce a

---

[92] http://www.fda.gov/AboutFDA/WhatWeDo/default.htm.

new device. Although the FDA often made my life difficult, I can attest that their requirements are well aligned with their purpose. FDA guidance documents make it clear that repeated, well-controlled, prospective interventional studies, and prespecified efforts to avoid bias, are expected as evidence of drug efficacy.[93,94] That sounds a lot like the six criteria of high-confidence science!

Broader than the FDA, the practice of medicine (i.e., guidelines on how to treat a specific condition) is increasingly built upon what is called "evidence-based medicine." Professional societies of physicians review the available evidence in their field of expertise and make a judgment: Should a drug, device, or procedure be recommended for a given condition? The foundation of evidence-based medicine is a hierarchical system for classifying evidence.[95] The levels of evidence were originally described in a 1979 report by the Canadian Task Force on the Periodic Health Examination.[96] They have been slightly modified since then to suit specific applications. Figure 13 is a simplified version for our discussion that makes use of a staircase metaphor for this hierarchy of scientific evidence; evidence level 1 is the highest, and 7 is the lowest.

This hierarchy of scientific evidence includes many technical terms for describing clinical studies. To make this more approachable, the following table defines these terms.

---

[93] US Department of Health and Human Services, Food and Drug Administration. Guidance for industry: Providing clinical evidence of effectiveness of human drug and biological products. May 1998. http://www.fda.gov/downloads/drugs/guidancecomplian ceregulatoryinformation/guidances/ucm072008.pdf.

[94] Katz, R. FDA: Evidentiary standards for drug development and approval. NeuroRx. 2004;1:307–16.

[95] http://www.ncbi.nlm.nih.gov/pmc/articles/PMC3124652/.

[96] Canadian Task Force on the Periodic Health Examination. The periodic health examination. Can Med Assoc J. 1979;121:1193–254.

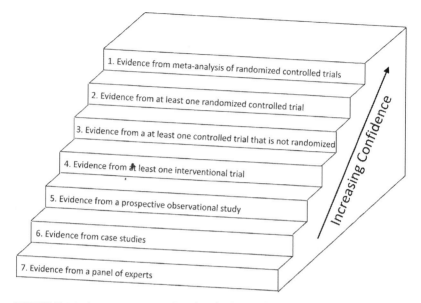

FIGURE 13: A staircase metaphor for the hierarchical levels of evidence from clinical trials.

Imagine that I have invented a new drug that I am convinced is the cure for cancer. I would like to sell this drug in the United States and to have the medical practice guidelines recommend that this drug be given to cancer patients. What will it take to convince the FDA and professional societies of physicians?

Level 7 (the lowest evidence level) is based on favorable opinions of a panel of experts. We often think of a panel of experts as being quite convincing—after all, they are the experts. Imagine that I spend some time with oncology experts discussing my new drug, and I succeed in convincing some of them (without any real data) that my drug is going to be the cure for cancer. I then get this group of converts together, we march into the offices of the FDA, and the experts offer up their opinions. Regardless of how many letters these experts list after their names, the most likely response from the

FDA would be: "Thanks, but please come back when you have some real data." (That is probably the nicest, most polite response they would offer.) Although I have well-educated experts on my side, all humans make subjective judgments and all humans can be wrong.

Term	Definition
Case study	A study conducted on a single individual.
Prospective	A study involving selecting a group in the present and following their outcomes into the future. The opposite is retrospective.
Interventional	A study where the researcher is able to control which subjects receive which treatment or intervention. The opposite is observational.
Controlled	A study where two groups are compared. One group receives a treatment or intervention while the other group (the control group) does not.
Randomized	A random number generator is used to determine which therapy is applied. This is a powerful way to remove bias by blocking the impact of confounding factors.
Meta-analysis	An analysis involving a combination of results from separate trials.

Level 6 is evidence obtained from case studies (i.e., results from individual patients). I could give my drug to one patient who has cancer and see what happens. If he or she doesn't die, I could race off to the offices of the FDA and tell them about how this one cancer patient didn't die: "You said to come back when I had some data; well, here it is!" Maybe I could also bring along my panel of experts to add fuel to the argument. Sorry, but this would not impress them. Perhaps this one patient wouldn't have died anyway, or maybe this one patient was quite unusual in responding to the drug. The FDA will require some repeatability—the 1st criterion of high-confidence science.

Level 5 consists of results from a prospective (not retrospective) observational study. This doesn't really apply to our example of an experimental cancer drug. A prospective observational drug study would involve watching what happens to people who have chosen to take the drug (for example, a drug that is available over-the-counter) and may include a comparison to people who have chosen not to take the drug. The decision to take the drug is not up to us; if it were up to us, it would be a prospective interventional study. However, my drug is not yet available on the market, so people cannot choose to take it or not take it. Although well-designed prospective observational studies can produce good results, confounding factors can introduce bias, because the study cannot control who takes what.

The FDA has this to say about the avoidance of bias in randomized trials versus the risk of bias in observational studies:

Whereas randomization in clinical trials prevents assignment of therapy based on prognosis, there is no such assurance of this kind of bias control in observational studies and registries that contain clinical outcomes. There are examples in the literature where the outcome from randomized clinical studies differs significantly from what had been reported in observational studies.[97]

97 FDA Guidance Document 1776: Design considerations for pivotal clinical investigations for medical devices—Guidance for industry, clinical investigators, institutional review boards and Food and Drug Administration Staff. November 7, 2013.

Essentially this says that because you don't control who does and does not get treatment in an observational study, there is a lot of room for bias, which leads to lower-confidence results (the 4th criterion of low-confidence science). For example, let's imagine that my cancer drug is freely available on the market for purchase, but it is very expensive, and patients have to pay for it themselves. If I conduct an observational study of 100 cancer patients who have chosen to take the drug versus 100 cancer patients who have not taken the drug, my results could be biased because of the high cost of the drug. Wealthy people who can afford the drug also have the resources to receive better care overall and therefore might have better outcomes than economically disadvantaged patients, regardless of the effects of my drug. The lack of control over who took the drug thus introduced a bias that may have confounded the results. In contrast, a study in which patients are randomized to receive the drug or a sugar pill would not have this bias. Therefore, the FDA prefers prospective interventional studies (the 3rd criterion of high-confidence science).

Notice that retrospective observational studies (such as those that search for King Tut's cause of death) do not appear anywhere on the hierarchical list of scientific evidence. That is because these studies offer such low confidence that they are only used to generate hypotheses for future prospective evaluations. As a medical example of a retrospective observational study, imagine that I have done some digging in a large database. The data was collected previously for another purpose, and I'm conveniently reusing it for my new study. The database included 50,000 subjects. Each subject entered the foods that he or she ate each day for a period of five years. The database also contains records of who died. By searching through this database (the practice known as "data mining"), I determine that subjects who consumed more than seven eggs per week had a slight but significantly increased risk of death compared with subjects who ate fewer eggs. My conclusion is that eating more than seven eggs per week is not good for you.

Perhaps you will remember that retrospective observational studies are not able to control variables and therefore can only suggest associations; they cannot conclude causation. Many biases could lurk behind my conclusion: maybe the egg-eaters were more likely to have dangerous careers like farming or construction. Maybe the egg-eaters were more likely to be smokers or heavy drinkers, which were the real causes of their early death. Rather than expecting this data to provide a causal relationship (i.e., eating eggs causes an increased risk of death), the low confidence in my finding implies that the finding should be treated as a preliminary result that will require further prospective evaluation. Retrospective observational studies are almost never considered for establishing evidence-based recommendations for the practice of medicine. If you face a life-or-death medical decision, you don't want to rely on a study like this to determine your fate. This is because retrospective observational studies are not repeatable (the 1st criterion of low-confidence science). Also, the results were previously recorded and not necessarily directly measured for the intended purpose (the 2nd criterion of low-confidence science). The study is also not prospective nor interventional (the 3rd criterion of low-confidence science), which leaves the door open for bias (the 4th criterion of low-confidence science). Finally, it requires assumptions (the 5th criterion of low-confidence science). On our hierarchical list of scientific evidence, we could include retrospective observational studies as "level 8" (although arguably it could be placed at level 7 or 6). Adding retrospective observational studies as level 8 results in the following figure.

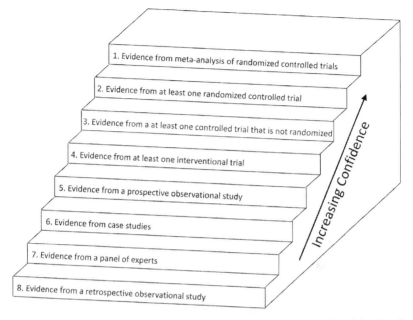

FIGURE 14: An extended staircase metaphor for the hierarchical levels of evidence from clinical trials.

Level 4 requires at least one interventional trial, meaning a trial where we control who gets treated. Here, I could design an experiment to treat 100 cancer patients with my drug. After two years of taking my drug, I find out that 90 of the original 100 patients are still alive. This seems like a great result! I run to the FDA and place the data at their feet. Still not good enough. They tell me that the study was not controlled, meaning that I cannot prove that my drug (i.e., the intervention) produced a better result than the current treatment method (i.e., a control group). In claiming that my drug helped in this study, I am making an implicit assumption (the 5th criterion of low-confidence science) that a control group would have had a worse outcome.

Level 3 requires a controlled trial that is not randomized. I then design a new experiment to study 200 patients with cancer: 100 will receive the drug, and 100 will be in the control group, receiving a placebo pill in place of the drug. Both groups will also receive the current best treatment. This is a true prospective interventional trial (the 3rd criterion of high-confidence science). After another two years of waiting, I find out that 80 patients in the control group remain alive, but 90 patients in the drug group remain alive. This is the most convincing data yet. When I present the data to the FDA, however, they ask a lot of questions about the two populations I studied. As it turns out, the control group included more cigarette smokers than were in the drug group. The FDA said that they could not be sure that the smoking hadn't led to more deaths in the control group (rather than my drug leading to fewer deaths in the treatment group). This is why randomization is so important in clinical trials: it blocks out confounding effects like cigarette smoking, which could create a bias in the results (the 4th criterion of low-confidence science).

Level 2 is what is known as the "gold standard" for clinical trials: the randomized controlled trial. In this trial, a random number generator decides if a given patient will receive the drug or the placebo. It is also best not to inform the patient or the medical staff about which treatment the patient receives. If the patient and/or medical staff know which treatment is being given, either or both parties might change their behavior and bias the results. This study design is now a "double blinded randomized controlled clinical trial." An even better situation is if the study is conducted at multiple clinical sites, because the particular practices at any one clinical site can add bias. This would be called a "double blinded multicenter randomized controlled clinical trial." When this trial is complete, the FDA will carefully review the study design and results. Many more factors will weigh into their decision. The FDA may accept this level 2 evidence and, if the study yielded positive results, they may approve my drug for sale in the United States.

Level 1 is one step better than level 2. A "meta-analysis" consists of results that are obtained from combining several separate trials. This is essentially repeating a randomized controlled trial to build confidence in the results (the 1st criterion of high-confidence science).

You may now be able to recognize that the established hierarchical levels of scientific evidence are well grounded in the six criteria of high-confidence science. For example, level 1 (meta-analysis of randomized controlled trials) involves repeated (1st criterion) prospective interventional (3rd criterion) studies. The results are directly measured (2nd criterion), and several aspects of the experimental design (randomization and placebo-control) are approaches to avoid bias (4th criterion). Figure 15 further compares this hierarchy of scientific evidence to the six criteria of high-confidence science.

The 1st criterion of high-confidence science (repeatability) is best demonstrated by a meta-analysis, which, as noted earlier, is a combination of results from several individual studies. One could also argue that trials conducted on multiple subjects practice repeatability when compared to trials on a single subject (i.e., case studies). The 2nd criterion (directly measurable and accurate results) is not accomplished by a panel of experts (level 7), because a panel of experts represents an opinion, not data. The 2nd criterion is unlikely to be accomplished in a retrospective study (level 8), because the data was obtained before the study was conceived. The 3rd criterion (prospective and interventional) is clearly accomplished by levels 1–4 and 6, because these levels involve studies that are planned in advance and include

	Characteristics of High-Confidence Science				
	Repeatable	Directly Measurable	Prospective Interventional	Avoid Bias	Avoid Assumptions
1. Meta Analysis of Randomized Controlled Trials	✓	✓	✓	✓	✓
2. Randomized Controlled Trial		✓	✓	✓	✓
3. Controlled Trial		✓	✓		✓
4. Interventional Trial		✓	✓		
5. Prospective Observational Study		✓			
6. Case Study		✓	✓		
7. Panel of Experts					
8. Retrospective Observational Study					

(Left axis label: Hierarchy of Scientific Evidence)

FIGURE 15: The criteria of high-confidence science are increasingly practiced within the higher levels of the well-established hierarchy of scientific evidence.

a treatment or intervention. The assessment of bias (4th criterion) is not nearly as black and white. Bias can enter a study in many different ways. Moving up the table from level 8 to level 1 is a general progression in the avoidance of bias. Because randomization is a powerful weapon against bias, levels 1 and 2 clearly have the lowest bias. Assumptions (5th criterion) can also enter a study in many different ways, but powerful assumptions are *required* in an interventional trial (level 4). Here, because there is no control group, one must assume an outcome for those who don't get the intervention in order to claim a benefit for those who do get the intervention. Powerful assumptions are also required in: level 5 (prospective observational study), because you must assume that any uncontrolled variables do not influence the result; level 6 (case study), because you must assume that the result on a single patient would apply to all patients; level 7 (panel of experts), because you must assume that the experts are correct; and level 8 (retrospective observational study), because you must assume that 1) no uncontrolled variables influence the result and that 2) any observed associations imply causality.

You may have noticed that the 6th criterion (sober judgment of results) is not listed, because this criterion is not related to the study design—it is related to how the results are summarized. Any statement about the benefit of a medical product (e.g., "this drug is guaranteed to grow hair on a bowling ball") is a "claim" in the eyes of the FDA. When a medical product is reviewed by the FDA, they will review the intended claims of the manufacturer to ensure that each claim is supported by the evidence: in short, to ensure that the manufacturer is practicing sober judgment of the results (6th criterion of high confidence). Likewise, before a professional society of physicians will recommend a drug, device, or procedure, they will review the conclusions of the published studies and determine if sufficient supporting data exists.

In summary, the well-established hierarchy of scientific evidence in the field of medicine validates the six criteria of high-confidence science. Compared with the application of science to determining the cause of King Tut's death, I hope that you can appreciate how the critical importance of good medicine places intense pressure on obtaining high-confidence evidence. If someone is wrong about the death of King Tut, no one gets hurt. If someone is wrong about the safety of a drug that you are taking, you and millions of others might be in big trouble.

Although the field of medicine places a particular emphasis on high-confidence evidence, the criteria of high-confidence science are certainly not lost on other fields of science. The criteria were largely described in chapter 2 by the example of dropping a bowling ball—a simple physics experiment. As another example, the concept of preregistration of studies is gaining momentum as a way to improve repeatability (1st criterion), reducing bias (4th criterion), and encouraging prospective study design (3rd criterion). This concept involves registering a study online *before* the study is conducted. The preregistration includes a description of the study and what will be analyzed. When the study is subsequently conducted, the information reported in the preregistration provides a type of scientific accountability: it removes the temptation to bury negative results (because people now know that the study was conducted) or to put a positive spin on disappointing results (because it forces the experimenters to stick to the analysis plan that they disclosed earlier). Positive results are generally rewarded with publications and future grants, which creates pressure to obtain positive results from every study. Conversely, negative findings have scientific value but are often cast aside because they generally do not help in obtaining publications and future grants. In 2015, the Center for Open Science began a program to offer $1 million to encourage the preregistration of studies in a very wide array of scientific disciplines.[98] This program was prompted by findings of poor repeatability among psychology experiments.[99] Although 97 out of 100 published studies reported positive results, the efforts to repeat each study led to positive results in only 36 of the 100 studies. Why the difference? The investigators who attempted to repeat the studies had no pressure (or bias) to obtain positive results (although perhaps

---

[98] Visit www.cos.io/prereg to find out more.

[99] Open Science Collaboration. Estimating the reproducibility of psychological science. Science. 2015;349:aac4716. DOI: 10.1126/science.aac4716.

they had the opposite: a desire to prove their point that many studies are not repeatable).

What about fields of science that cannot practice high-confidence science? Interventional trials are not possible or not ethical in plenty of fields of science. Qualitative, historical, and quasi-experimental fields of study are common. For example, if you wanted to study the impact of living near a nuclear power plant on standardized test scores (like the ACT or SAT), you cannot ethically conduct a randomized controlled trial, because you cannot randomly assign some people to move to the neighborhood of a nuclear power plant and others to move far away. The highest-confidence study you could perform to address this kind of question is level 5 (prospective observational study). If you wanted to study the cause of death of King Tut, the highest-confidence study you could perform is level 8 (retrospective observational study). In these situations, I am not suggesting that these scientists are not good scientists; I'm saying that they have chosen (perhaps unknowingly) to try to answer a question that is difficult to answer using the tool of science. Arriving at the truth becomes akin to refereeing a football game while looking through a drinking straw.

Appendix B

Anticipated Objections

Without a doubt, portions of this book will be objectionable to some readers. The following is an attempt to address some of these objections. In each case, the "**O**" is the anticipated objection, and the "**R**" is the response. To view additional material or to post a comment or question, please visit **www.scientificevolution.com** or contact the author at **scientificevolution.com@gmail.com**.

1. **O:** Your work is very offensive to paleontologists and paleoanthropologists—you discredit and disregard their work and claim that they are bad scientists.

    R: I've been very careful in my choice of words to avoid judgment of "good" or "bad." What I have said is that these scientists have chosen to work in a field where evidence is very sparse and truth is extremely difficult to determine. Despite their great intelligence, impressive educational credentials, admirable persistence, and careful technique, the truth may remain elusive to them, as we saw in the study of King Tut's death. Bias and assumptions are very powerful forces in this type of science, such that the influence of bias and assumptions can overwhelm the influence of the sparse data. Because practicing the six criteria of high-confidence science is so difficult in these fields, the temptation is to assume that low-confidence evidence (the best and only evidence that is available) should be treated as high-confidence evidence. But this would be a misrepresentation of the evidence. These fields of study can make contributions that are higher in confidence—such as describing the location, size, and distribution of fossils; studying the chemical composition of fossils; etc.

2. **O:** Your criteria for "high-confidence" and "low-confidence" science apply only to medical research. The scientific method has a fundamentally different approach for medicine than it does for evolution. You cannot force the scientific method that applies to one area onto another field of study.

R: I completely agree that randomized placebo-controlled clinical trials do not apply to the study of evolution; they are more specific to the field of medicine. The practice of high-confidence science, however, applies to all fields of science. The six criteria of high-confidence science were explained through the example of dropping a bowling ball, which has little to do with medicine. Chapter 7 makes it clear that evolution can be studied with high confidence. The examples of chapter 7 clearly demonstrate experimental evolution that is repeatable, directly measurable, prospective, and interventional, with minimal bias or assumptions, and with results that are presented with sober judgment. Can any field of science argue that it is better if a result cannot be repeated? Can any field of science argue that indirect measurement is preferred over direct measurement, or that retrospective observational study (with lack of control over confounding variables) is preferred? Or that bias is good for uncovering the truth? That making lots of assumptions is the best approach to science? That overstating confidence or the scope of results is the best way to explain a finding? The six criteria are fundamental tenets of any application of the tool of science.

3. **O:** While arguing that the fossil record represents low-confidence evidence, you focus on the individual fossils. The evidence is much more diverse than just fossils—the layers they are buried in and what they contain, the estimates of ages, the surrounding layers, etc., give a much more complete picture.

    R: All of the evidence contained in the fossil record, when applied to the study of macroevolution, is low-confidence science. The putative macroevolutionary processes cannot be repeated. Macroevolution cannot be studied prospectively. It cannot be studied directly. The widely acknowledged limitations of the fossil record require many assumptions for interpretation, and bias is able to overwhelm the influence of the available evidence.

4. **O:** A very large quantity of low-confidence evidence (for example, the entirety of the fossil record) should add up to more than a few high-confidence experiments.

    R: As described in chapter 3, the use of the drug digoxin has been studied with predominantly low-confidence approaches among more than 300 thousand patients (i.e., a very large quantity of low-confidence evidence). These studies have led to the widespread adoption of the drug. Yet recent meta-analyses strongly suggest that the drug is responsible for increased mortality. In short, *quantity* of low-confidence evidence does not translate to increased *quality* of evidence.

5. O: Your work is a simple repackaging of the creationist arguments for historical science versus experimental science, which have been refuted: for example, the philosopher of science Carol Cleland's article "Historical Science, Experimental Science, and the Scientific Method" (Geology. 2001: 29; 987-990).

R: The argument for historical science versus experimental science only touches upon the 3rd criterion (prospective interventional study versus retrospective observational study). The other five criteria of high-confidence science are not mentioned in this argument. Cleland's article is an attempt to dismiss the 3rd criterion. Cleland argues that retrospective study is no less credible than prospective study by emphasizing a philosophical nuance rather than scientific pragmatism. Cleland relies on the philosophical nuance that it is impossible to control every possible confounding factor in a prospective experiment. She then claims that this uncertainty in prospective experimentation is equivalent to the uncertainty of conclusively establishing the occurrence of a hypothesized past event. She is essentially claiming that nothing can be proven with absolute certainty and that if there are no absolute certainties, then there can be no differences between levels of confidence in prospective versus retrospective studies. I have chosen to use the terms "high confidence" and "low confidence" rather than implying absolute certainty, because I recognize that nothing can be proven with absolute certainty. However, we cannot allow the absence of absolute certainty to cause us to abandon our appreciation for the levels of confidence. Returning to our focus on the pragmatic practice of science, I'm grateful that Cleland's assertions that retrospective and prospective science are equally valid had no influence on the chemotherapy regimen for my friend's cancer treatment. I shudder to think of what would happen to medical practice if her philosophy were to be adopted!

6. **O:** Lenski's experiment could have produced entirely new genes and complex new proteins, but these results simply haven't been reported yet.

R: I agree that this is a possibility, but until this is clearly demonstrated, it certainly wouldn't count as high-confidence evidence. I look forward to additional results from Lenski's study and other high-confidence prospective studies of evolution—this is the right approach to uncover the truth.

7. **O:** Your central argument is that high-confidence experimental studies of evolution (e.g., Lenski's *E. coli*, malaria, and tryptophan synthase) failed to produce significant new genetic material, but there is significant new genetic material between humans and chimpanzees. Therefore, the evolution of humans from a common ancestor is not supported. However, the experimental studies of evolution occurred under very specific, controlled conditions. The conditions that produced humans from our common ancestor were dramatically different and have not been tested in a laboratory.

R: This argument demonstrates subordination of science to faith. As described in chapter 8, the experimental conditions for the high-confidence studies of evolution applied strong and consistent selective pressures over thousands of generations. In contrast, nature rarely (if ever) provides such strong and consistent selective pressures over the timespans required for macroevolution.

Although formation of significant new genetic material from random mutations and natural selection has never been demonstrated, some will choose to believe that it is still possible. One definition of faith is "confidence in what we hope for and assurance about what we do not see." Despite what high-confidence science is telling us, some will refuse to turn away from their faith in evolution. American philosopher Thomas Nagel refers to the current paradigm of evolutionary biology as "a heroic triumph of ideological theory over common sense." I would like to amend this quote: "Grand evolution is a heroic triumph of ideological theory over common sense and high-confidence science."

8. **O:** You make a big deal about the boundary between microevolution and macroevolution, claiming that all microevolution is consistent with both sides of the debate and is supported by high-confidence evidence but that macroevolution is not supported by high-confidence evidence. But you don't provide a clear definition of where microevolution ends and macroevolution begins. Without a clear boundary, you can simply redefine the imaginary boundary when new high-confidence evidence of larger-scale evolution appears.

R: The most convenient boundary between microevolution and macroevolution is a taxonomic level like species, genus, or family, but we must recognize that our taxonomic classification system is not perfect. New information commonly results in modifications to classifications (after some debate). For example, scientists recently discovered that dromedary camels and llamas can reproduce, making a "cama." Camels and llamas currently share the same taxonomic family (Camilidae), but they each have a different genus and species. If we find that camas can reproduce with one another and with llamas and camels, this would raise an argument that camels and llamas are the same genus and species, according to the "biological species" concept, which is one of many definitions of "species." Defining where microevolution ends and macroevolution begins according to a threshold taxonomic level (like species or genus or family) may be convenient, but it will never be as clearly defined as we would like. The subjectivity in the boundary between micro- and macroevolution can be no better than the subjectivity in the classification of organisms.

9. **O:** Your claims that abiogenesis and eukaryogenesis are not scientific show that you are eager to give up on science and bow to the "god of the gaps" (a belief that gaps in scientific knowledge must point to the existence of God). Just because science hasn't yet explained something doesn't mean that science will not explain it in the future.

   R: I've seen many attempts to explain abiogenesis with science. Each of them attempts to provide enough hints of the possibility that the "leap of faith" required to believe in abiogenesis becomes acceptable to those who want to believe. Koonin's view (from chapter 9) is refreshing:

   > The origin of life is the most difficult problem that faces evolutionary biology and, arguably, biology in general. Indeed, the problem is so hard and the current state of the art seems so frustrating that some researchers prefer to dismiss the entire issue as being outside the scientific

domain altogether, on the grounds that unique events are not conducive to scientific study.[100]

The fundamental problem lies in the attempt to study a unique event that supposedly happened 4 billion years ago. While I have no doubts that future claims will be made for evidence of abiogenesis or laboratory demonstration of abiogenesis, we will never have confidence in a putative original abiogenesis event. More fundamentally, abiogenesis is not falsifiable and is therefore pseudoscience.

10. **O:** You claim that each species has 10–20 percent orphan genes. You then claim that macroevolution is not supported (in the case of humans evolving from our last common ancestor), because high-confidence evidence for evolution does not demonstrate the production of novel genes. But you also claim that speciation occurs and that it is directly observable. This is self-contradictory.

R: The claims of 10–20 percent orphan genes in each species will certainly have exceptions. The observable examples of speciation (chapter 7) involved a simple combination of the genetic material of the parents. There are no orphan genes in these exceptional cases. There will undoubtedly be other exceptions where speciation has occurred with minimal modification of genetic material.

11. **O:** You argue that the extrapolation of the variety of dog breeds to millions of years (to support macroevolution) is not scientific, but then you extrapolate lessons learned from microbes

---

[100] Koonin, EV. The logic of chance: The nature and origin of biological evolution. Upper Saddle River, NJ: Pearson Education, Inc; 2012, p. 351.

(the high-confidence evidence of evolution in chapter 7) to discredit the evolution of humans from primates. This is hypocritical.

R: There are a few major differences between the dog argument for macroevolution and my human argument against macroevolution. First, the dog argument is a clear extrapolation over time: dog observations over a short time are extrapolated to expected dog evolution over millions of years. In contrast, my argument against human evolution involves taking the lessons learned from observing thousands of generations and trillions of organisms (microbes) and applying them to thousands of generations and roughly one trillion hominids. Although the expected time span of human evolution is much greater than the time span of observed microbial evolution, it is the number of generations and the number of organisms that matter to evolution, not the number of years. My human argument extrapolates over species, whereas the dog argument extrapolates over time. The microbes we studied in chapter 7 included both prokaryotes and eukaryotes: vastly different organisms that both provide similarly constrained views of evolution. Extrapolating this to humans will certainly not be perfect, but it is more tenable than extrapolating a short-term observation to an expectation over millions of years. Second, the dog argument is based on phenotype, not genotype. We look at dog breeds and appreciate the large phenotypic variety. I have not seen this phenotypic variation mapped to genotypic variation. In other words, how significant are the genetic differences between breeds? All dog breeds are the same species, so the genetic differences between breeds are likely very minor: probably as minor as the variations that Lenski found in his *E. coli*. Genetics are the basis of evolution, so that is where we need to take our arguments.

If the genetic differences between dog breeds are minimal, one cannot extrapolate this over time to assume that the large-scale genetic innovations required for macroevolution would occur. The human argument of chapter 8 is based on genotype, whereas the dog argument is based on phenotype. My human argument does have its limitations, which I state at the end of chapter 8.

12. O: Your book is an underhanded approach to bring religion into public schools.

R: Chapter 10 contains my suggestions for changes to public school curricula. I am promoting the teaching of the criteria of high-confidence science and the teaching of evolution with a critical and sober appraisal of the evidence, similarly to the way it is presented in this book. Children need to learn that microevolution is well supported by high-confidence evidence. In contrast, macroevolution is far from proven. As discussed in chapter 8, current high-confidence science does not support macroevolution. Young people should also learn that science is a tool that cannot answer every question; the simple question of "How did life begin?" is one of them. By teaching abiogenesis and eukaryogenesis, public schools promote a faith-based worldview. Claiming that I am promoting the teaching of religion in public schools is therefore rather hypocritical.

Bibliography

Andriantsoanirina, V. Chloroquine clinical failures in *P. falciparum* malaria are associated with mutant Pfmdr-1, not Pfcrt in Madagascar. PLoS ONE. 2010;5;e13281.

Belk, C, Borden Maier, V. *Biology: Science for life (with Physiology).* Third edition. San Francisco: Benjamin Cummings; 2010.

Blount, ZD, et al. Historical contingency and the evolution of a key innovation in an experimental population of *Escherichia coli*. PNAS. 2008;105:7899–906.

Blount, ZD, et al. Genomic analysis of a key innovation in an experimental *Escherichia coli* population. Nature. 2012;489:513–20.

Burke, MK, et al. Genome-wide analysis of a long-term evolution experiment with *Drosophila*. Nature. 2010:467;587–90.

Canadian Task Force on the Periodic Health Examination. *The periodic health examination.* Can Med Assoc J. 1979;121:1193–254.

Carlton, BC, and Brown, BJ. Gene mutation. In: Gerhardt, P, ed. *Manual of Methods for General Bacteriology.* Washington, DC: American Society for Microbiology; 1981: 222–42.

The Chimpanzee Sequencing and Analysis Consortium. Initial sequence of the chimpanzee genome and comparison with the human genome. Nature. 2005:437;69–87.

Cleland, C. Historical science, experimental science, and the scientific method. *Geology*, November, 2001;29:987-90.

Collins, FS. *The language of God: A scientist presents evidence for belief.* New York: Free Press; 2006.

Cooper, VS, et al. Mechanisms causing rapid and parallel losses of ribose catabolism in evolving populations of *Escherichia coli*. J Bacteriol. 2001;183:2834–41.

Coyne, JA. *Why evolution is true.* London: Penguin Books; 2009.

D'Alessandro, U, and Buttiens, H. History and importance of antimalarial drug resistance. Tropical Medicine and International Health. 2001:6;845–8.

Dawkins, R. *The greatest show on Earth.* New York: Free Press; 2009.

Demuth, JP, et al. The evolution of mammalian gene families. PLoS One. 2006;1:e85.

Dines, JP. Sexual selection targets cetacean pelvic bones. Evolution. 2014;68:3296–306.

Ebersberger, I, et al. Mapping Human Genetic Ancestry. Mol Biol Evol. 2007;4:2266–76.

Ecker, A. PfCRT and its role in antimalarial drug resistance. Trends in Parasitology. 2012;28:504–14.

FDA Guidance Document 1776: *Design considerations for pivotal clinical investigations for medical devices—Guidance for industry, clinical investigators, institutional review boards and Food and Drug Administration Staff.* November 7, 2013.

Fidock, DA, et al. Mutations in the *P. falciparum* digestive vacuole transmembrane protein PfCRT and evidence for their role in chloroquine resistance. Molecular Cell. 2000;6:861–71.

Garg, R, et al. The effect of digoxin on mortality and morbidity in patients with heart failure. N Engl J Med. 1997;336:525–33.

Gauger, AK, et al. Reductive evolution can prevent populations from taking simple adaptive paths to high fitness. BIO-Complexity. 2010;2:1–9.

Glynn, RJ, et al. Effect of low-dose aspirin on the occurrence of venous thromboembolism. Ann Intern Med. 2007;147:525–33.

Hall, BG. Chromosomal mutation for citrate utilization by *Escherichia coli* K-12. J Bacteriol. 1982;151:269–73.

Harrison, RG. Postmortem on two pharaohs: Was Tutankhamun's skull fractured? Buried Hist. 1971;4:114–29.

Hawass, Z, et al. Ancestry and pathology in King Tutankhamun's family. JAMA. 2010;303:638–47.

Hughes, JF. Chimpanzee and human Y chromosomes are remarkably divergent in structure and gene content. Nature. 2010;463:536–9.

Hutchison, CA III, et al. Global transposon mutagenesis and a minimal mycoplasma genome. Science. 1999;286:2165–9.

Hyland, K. *Hedging in scientific research articles.* Amsterdam and Philadelphia: John Benjamins Publication Company; 1998. DOI: 10.1075/pbns.54.

Johanson, D, and Edey, M. Lucy: *The beginnings of humankind.* New York: Simon & Schuster; 1981.

Johanson, D, and Wong, K. *Lucy's legacy: The quest for human origins.* New York: Three Rivers Press; 2009.

Katz, R. FDA: Evidentiary standards for drug development and approval. NeuroRx. 2004;1:307–16.

Khalturin, K, et al. More than just orphans: Are taxonomically-restricted genes important in evolution? Trends in Genetics. 2009;25:404–13.

Koonin, EV. *The logic of chance: The nature and origin of biological evolution.* Upper Saddle River, NJ: Pearson Education, Inc.; 2012.

Kublin, JG, et al. Reemergence of chloroquine-sensitive *Plasmodium falciparum* malaria after cessation of chloroquine use in Malawi. J Infect Dis. 2003;187:1870–5.

Li, G-W, et al. The anti-Shine-Dalgarno sequence drives translational pausing and codon choice in bacteria. Nature. 2012:484;538–41.

Lütgens, M, and Gottschalk, G. Why a co-substrate is required for anaerobic growth of *Escherichia coli* on citrate. J Gen Microbiol. 1980;119:63–70.

Nye, B. *Undeniable: Evolution and the science of creation.* New York: Saint Martin's Press; 2014.

Open Science Collaboration. Estimating the reproducibility of psychological science. Science. 2015;349:aac4716. DOI: 10.1126/science.aac4716.

Pilcher, H. All alone. New Scientist. January 19, 2013;38–41.

Puhan, MA, et al. Discussing study limitations in reports of biomedical studies: The need for more transparency. Health and Quality of Life Outcomes. 2012;10:23.

Quandt, EM, et al. Recursive genomewide recombination and sequencing reveals a key refinement step in the evolution of a metabolic innovation in *Escherichia coli*. PNAS. 2014;111;2217–22.

Raven, PH, et al. *Biology. Eighth edition.* New York, NY: McGraw-Hill, Inc.; 2008.

Reece, JB, et al., eds. Campbell *Biology. Ninth edition.* Upper Saddle River, NJ: Pearson Education, Inc.; 2011.

Robinson, JG, et al. Safety and efficacy of alirocumab in reducing lipids and cardiovascular events. N Engl J Med. 2015;372:1489–99.

Sanford, JC. *Genetic entropy. Fourth edition.* FMS Publications; 2014.

Scheutz, F, and Strockbine, NA. In: Garrity, GM, et al., eds., *Bergey's Manual of Systematic Bacteriology, Volume 2: The Proteobacteria.* Springer; 2005: 607–24.

Sidhu, AB, et al. Chloroquine resistance in *Plasmodium falciparum* malaria parasites conferred by pfcrt mutations. Science. 2002;298:210–13.

Sniegowski, PD, Gerrish, PJ, and Lenski, RE. Evolution of high mutation rates in experimental populations of *Escherichia coli.* Nature. 1997;387:703–5.

US Department of Health and Human Services, Food and Drug Administration. *Guidance for industry: Providing clinical evidence of effectiveness of human drug and biological products.* May 1998.

Vamos, M, et al. Digoxin-associated mortality: A systematic review and meta-analysis of the literature. Eur Heart J. 2015;36:1831–8.

Wang, ZQ, et al. Digoxin is associated with increased all-cause mortality in patients with atrial fibrillation regardless of concomitant heart failure: A meta-analysis. J Cardiovasc Pharmacol. 2015;66:270–5.

Watson, JD, and Crick, FH. A structure for deoxyribose nucleic acid. Nature. 1953;171:737–8.

White, NJ. Delaying antimalarial resistance with combination chemotherapy. Parassitologica. 1999;41:301–8.

White, NJ. Antimalarial drug resistance. J Clin Invest. 2004;113:1084–92.

Wu, D-D, et al. De novo origin of human protein-coding genes. PLoS Genet. 2011;7:e1002379.

Wu, Y, et al. Transformation of *Plasmodium falciparum* malaria parasites by homologous integration of plasmids that confer resistance to pyrimethamine. Proc Natl Acad Sci. 1996;93:1130–4.

Zhang, YE, et al. New genes contribute to genetic and phenotypic novelties in human evolution. Current Opinion in Genetics & Development. 2014;29:90–6.

Index

abiogenesis   **136-147**, 150, 152, 181, 182, 184

accuracy   **9-12**, 134

alirocumab   26-29

amino acid   **86-87**, 89, 106, 107, 110, 111, 112, 116, 132, 140-142

ancestry   60-63, 129, 130, 155, 157, 179, 182

Answers In Genesis   4, 44, 78

antibiotic   40, 41, 43, 106, 108, 109, 154

appendix, human   63, 64, 71, 72, 74

Archaea   102, 144, 145

artificial selection   81, 84, 95, 109

aspirin   20-21

association   **14-15**, 29, 34, 36, 37, 49-51, 61, 166, 171

assumption   10, 13, 15, 16, **18-19,** 23, 24, 40, 43, 44, 45, 53, 54, 57, 61, 64, 65, 68, 71, 73, 74, 77, 78, 80, 81, 83, 84, 93, 108, 126, 134, 150-152, 156, 166, 167, 170-171, 175-177, 184

atheism   4, 61, 152

atovaquone   108

atrial fibrillation   31-33

Australopithecus afarensis   58, 60, 62

bacteria   40, 86, 95-98, 102, 144, 145

base pair   85, 86, 89, 97, 116, 127, 131, 132, 134

bias   10, **15-17**, 19, 23, 24, 28, 34, 50, 53, 57, 59, 61, 63, 68, 71, 74, 80, 81, 83, 93, 94, 108, 126, 148, 150, 151, 154, 161, 164-166, 168-170, 172, 175-177

biogeography	74-79, 92
BioLogos	5
biology	1, 3, 42, 44, 48, 50, 68, 74, 137, 139, 142, 146, 150-153, 155, 180, 181
blood clot	12, 20
blood pressure	11-12
bowling ball	8-15, 18-20, 140, 171, 172, 176
Burke, Molly	93
calibration	10, 11
cama	180-181
camel	180-181
cancer	16-17, 162-165, 167, 168, 179
Canis familiaris (see dog)	
case study	162, 163, 167, 171
causality	14, 15, 49, 50, 51, 61, 171
cell nucleus	102, 145
Center for Open Science	172
chimpanzee	58, 121, 122, 125-131, 133, 134, 147, 179
chirality	140
chloroquine	102, 104-110, 117, 121, 124, 132, 134
cholesterol	8, 22, 23, 25-28, 31, 32, 37, 47, 80, 95, 102, 119
chromosome	92, 93, 129, 130, 134
citrate	98, 100, 101, 105, 117, 121
Collins, Francis	5, 127, 147
common ancestor	40, 44, 66-69, 121, 122, 125, 126, 129, 131-134, 147
confounding factors	14, 15, 18-20, 23, 24, 34, 164, 168, 176, 178

control, experimental	10, 13, 14, 15, 18-20, 23, 24, 34, 95, 110, 166, 178, 179
controlled study	31, 32, 94, 119, 120, 167-169, 173, 173
Coyne, Jerry	77, 82, 83, 92, 131, 139
creation (creationism)	1, 3, 4, 44, 53, 57, 64, 68, 74, 77, 78, 99, 139, 146-148, 178
Crick, Francis	66, 85
curriculum	150-153, 155
Darwin, Charles	1, 47, 66, 76, 77, 85, 90, 136, 137
Dawkins, Richard	3, 6, 43, 66, 68, 69, 77, 82-84, 99, 101, 120, 144
design	4, 67-75, 80, 99
digoxin (digitalis)	31-33, 177
dinosaur	37, 43, 48, 52, 58, 92
DNA	36, 65-67, **84-90**, 96, 97, 101, 102, 107, 108, 110, 113, 114, 115, 125-128, 131, 132, 134, 141, 142, 145
dog	40, 41, 43, 56, 81-84, 91, 92, 117, 120, 182-184
domain	102, 144, 145
drinking straw	33, 36, 47, 56, 58, 173
Drosophila melanogaster (see fruit fly)	
education	14, 150, 154, 175
embryology	1, 71
ENCODE study	65, 127
endosymbiosis	146
enzyme	87, 105, 110-114, 116, 132
Escherichia coli (E. coli)	86, 95-102, 109, 111, 112, 115-118, 121, 122, 125, 179, 183
Eukarya (eukaryotes)	68, 102, 128, 135, 144-146, 183

eukaryogenesis — 145-147, 150, 181, 184

exponential notation — 90-91

faith — 4-6, 139, **140**, 142-148, 154, 156, 157, 179-181

falsifiability — **139**, 140, 142, 143, 146, 152, 182

finches, Darwin's — 43, 44, 76

fitness — 115, 116, 118, 132, 133

Food and Drug Administration (FDA) — 16, 17, 26, 56, 119, 120, 160-169

fossil — 1, 37, **46-60**, 62, 65, 69, 80, 83, 92, 102, 148, 154, 156, 176, 177

fruit fly — 93-95, 109, 117, 127, 131

Gallup poll — 1, 2, 5, 136

Gauger, Ann — 111-113, 121, 123, 124

gene — 67, 68, **86**, 90, 93, 99-101, 110-118, 120, 123, 128-134, 141, 142, 179, 182

generalized evolution — **40-42**, 93, 97, 98, 104, 110, 118

genetic entropy — 64, 74

genetics — 1, 36, 65, 66, 85, 93, 94, 98, 99, 102, 105, 107, 108, 110, 117, 118, 128, 131, 154, 179, 180, 182-184

genotype — 66, 69, 183, 184

glucose — 96-98

Google — 7, 34, 57, 58, 62, 122, 155

grand evolution — **40-45**, 46, 47, 58, 65, 117, 136, 144-148, 151, 152, 155, 180

Gurche, John — 156, 158

Haeckel, Ernst — 71

heart attack — 25, 28

heart failure — 31-33

hedging	21, 22, 37, 57
hemoglobin	104-105
homology	66-71, 73, 80, 128, 129
honeycreepers, Hawaiian	76
Horner, Jack	52, 58, 92
Human Genome Project	5, 65, 126, 127
interventional	**13-14**, 16, 24, 49, 50, 57, 81, 83, 93, 95, 107, 108, 110, 150, 161, 163-170, 173, 176, 178
irreducible complexity	99, 145
islands, continental	76
islands, Galapagos	76-77
islands, oceanic	76-78
Johanson, Donald	58-62
junk DNA	65, 126, 127
Jurassic Park	48
King Tut, see Tutankhamun	
Koonin, Eugene	137, 145, 181
last common ancestor	121, 122, 125, 126, 131-134, 147, 182
Lenski, Richard	95, 97, 98, 100-102, 105, 111, 113, 117, 120, 121, 123, 124, 179, 183
limitations of science	6, 7, 12, 37, 56, 143, 147-149, 153, 154
limitations, study	22, 29, 57, 134, 155
Linnaeus, Carl	41, 66
llama	180-181
Lucy	57-63, 155-158
macroevolution	**41-43**, 45-47, 49, 54, 56, 57, 63-66, 69-71, 73-75, 78-83, 110, 118-122, 131, 133, 148-151, 154, 177, 179-184

malaria	36, 86, **102-110**, 117, 121, 123-125, 132, 134, 179
mammal	46, 50, 54, 74, 76, 77
mammal, marsupial	67
mammal, placental	67
media	4, 155
medicine	12, 24, 44, 119, 151, 159-161, 166, 171, 176
Mendel, Gregor	85, 86
meta-analysis	163, 169
metabolism	87, 98, 100, 101, 141
Michelangelo	156
microevolution	**41-45**, 56, 78, 82, 84, 118, 133, 148, 149, 151, 154, 180, 181, 184
Miller, Stanley	140, 142
mitochondria	146
molecular biology	1, 84, 137
mosquito	104
museum	4, 61, 62, 63, 155-158
mutation	41, 74, **89-90**, 96-98, 100, 101, 107-111, 116-123, 128, 132, 133, 180
Mycoplasma genitalium	141
Nagel, Thomas	180
natural selection	41, 82, 84, 85, 90, 94, 96, 99, 111, 121, 180
naturalism	152
nucleotide	**85-90**, 96, 110, 112
nucleus	102, 145
Nye, Bill	3, 4, 43, 44, 67-69, 143

observational	**13-16**, 20, 24, 31, 34, 37, 50, 57, 61, 83, 105, 107, 108, 119, 120, 164-173, 176, 178
organelle	102, 105, 145, 146
orphan gene	68, **128-130**, 182
oxygen	98, 100, 101, 105, 121
paleoanthropology	58, 59, 175
paleontology	52, 53, 56, 155, 175
parasite	36, 86, **103-105**, 108, 117
PfCRT	106-108
phenotype	**66**, 69, 183, 184
Pilcher, Helen	128
placebo	20, 26, 27, 28, 31, 168, 169, 176
Plasmodium falciparum (see malaria)	
preregistration	172
prokaryote	68, 102, 135, 144-146, 183
promoter	**86**, 100, 118
prospective	**13-14**, 16, 19, 23, 24, 27, 34, 49, 50, 57, 80, 81, 83, 84, 93, 94, 95, 105, 107, 110, 117, 121, 133, 150, 161, 164-173, 176-179
protein	65, **86-90**, 101, 106, 108, 110, 112, 118, 120, 126, 128-130, 136, 137, 140-142, 179,
public school	1, 4, 150-155, 184
pyrimethamine	109
Quandt, Erik	100
quinine	104
randomization	27, 164, 165, **168**, 170

randomized controlled clinical trial	31, 32, 119, 120, 164, **168**, 169, 173, 176
recombinant DNA	114-115
reconciliation	6, 78, 147-149
recurrent laryngeal nerve	71-72
red blood cells	104
religion	3-5, 15, 144, 151, 152, 154, 184
repeatability	9-10, 11, 16, 23, 24, 47, 48, 50, 57, 60, 80, 81, 83, 93, 96, 107, 113, 150, 161, 164, 166, 169, 172, 173, 176, 177
retina	72-73
retrospective	**13-15**, 16, 20, 24, 34, 37, 49, 50, 57, 61, 80, 119, 120, 165, 166, 169, 171, 173, 176, 178
RNA	86-89, 97, 141, 142
scientific notation	90-91
scope, experimental	**19-21**, 24, 28, 29, 61, 84, 93, 105, 108, 110, 120-125, 177
sedimentary layers	46, 49, 51, 53
separation of church and state	151-152
Smithsonian Museum	61-63, 158
snake oil salesman	159
sober judgment	19-23, 28, 54, 81, 84, 93, 99, 108, 134, 146, 151, 155, 171, 176
speciation	**42**, 44, 45, 56, 76, 78, 91-93, 117, 182
statin	25, 27
succinate	100
supernatural	53, 139, 151-153

taxonomically-restricted gene (see orphan gene)

taxonomy   42, 66, 144, 148, 180, 181

TED talk   52, 58

transcription   86

transitional fossil   46-47, 51, 54, 148, 154,

translation   86-90

tryptophan   110-116, 118, 121, 132, 179,

Tutankhamun   8, 33-39, 47, 48, 102, 104, 165, 171,
                                173, 175

Urey, Harold   140, 142

vestigial organs   63-65, 72, 80

Watson, James   66, 85

Welsh groundsel   92-93

whale evolution   47, 50, 63, 64

wolf   67, 81

worldview   2-5, 144, 146, 147, 184

YouTube   3, 13

To view additional material or to post a comment,
please visit **www.scientificevolution.com**

or send comments to **scientificevolution.com@gmail.com**

Made in the USA
San Bernardino, CA
19 December 2016